Introduction

This workbook is designed to be used in conjunction with *The Essentials of AQA Maths – Linear Specification A (3301): Intermediate Tier*. It is matched page for page to this revision guide and consists of structured questions with spaces for answers, plus extension questions.

The contributors to this workbook are former or current GCSE examiners who have drawn on their experience to produce interesting and challenging exam-style questions. The worksheets are designed to reinforce understanding of the material in the revision guide, which is covered in the specification.

Details of our other AQA Maths Revision Guides and Workbooks can be found on the inside back cover.

Consultant Editor:

John Proctor B.Sc. (Hons), Cert. Ed.
Director of Specialist College,
St. Mary's Catholic High School,
Astley,
Manchester
(An 11-18 Specialist Mathematics
and Computing College)

Editor:

Kay Chawner
Formerly a mathematics teacher
at The Ridgeway School, Swindon.

*Although this workbook is intended for candidates following the linear specification, it can also be used by candidates following the MODULAR B (3302) specification.

Contents

Place Value

1 Complete the following table. The first row has been done for you.

	Place Value of Digits					Number
	10 000 Ten Thousands	1 000 Thousands	100 Hundreds	10 Tens	1 Units	
			3	2	6	Three hundred and twenty six
a)		5	3	0	7	....
b)	7	3	1	5	8	....
c)			....	....	....	Five hundred and sixty
d)		1	....	....	7	.... , two hundred and forty ....
e)	....	....	....	....	....	Fourteen thousand and fifty two

2 Round 493 507 to the nearest…

a) 10 000 .... b) 1 000 .... c) 100 .... d) 10 ....

3 Round 109 109 to the nearest…

a) 10 000 .... b) 1 000 .... c) 100 .... d) 10 ....

4 If the attendance at a football match was 32 000 to the nearest thousand, what is the lowest possible number and the highest possible number of people that could have attended the match?

Lowest possible number: .... Highest possible number: ....

5 The number of boys who attended school on a particular day was 440 to the nearest ten. On the same day the number of girls who attended school was 460 to the nearest ten.

a) What is the lowest possible difference between the number of boys and the number of girls who attended school on that day? ....

....

b) What is the highest possible difference between the number of boys and the number of girls who attended? ....

....

6 Draw a table to show the place value of each digit in the following numbers:
a) one b) eighty seven c) four hundred and nine d) six thousand, two hundred and twenty three e) fifty thousand and five

7 Round 30 715 to the nearest.... a) thousand b) hundred c) ten

8 Write the equivalent value of half a million in... a) numbers b) words

1 **Here are eight numbers:**

3   4   6   7   11   15   20   21

a) Which three numbers are even numbers? ....

b) Which two numbers are factors of 40? ....

c) Which two numbers are factors of 45? ....

d) Which four numbers are factors of 42? ....

e) Which two numbers are multiples of 7? ....

f) Which three numbers are multiples of 2? ....

g) Which three numbers are prime numbers? ....

h) Which number has an odd number of factors? ....

2 **Here are ten numbers:**

5   8   11   19   22   24   31   36   47   81

a) Which six numbers are odd numbers? ....

b) Which three numbers are factors of 72? ....

c) Which three numbers are factors of 110? ....

d) Which three numbers are multiples of 3? ....

e) Which three numbers are multiples of 4? ....

f) Which five numbers are prime numbers? ....

g) Which two numbers have an odd number of factors? ....

3 **What is the reciprocal of...**

a) 8? .... **b)** 25? .... **c)** 0.5? .... **d)** $\frac{3}{4}$? ....

4 **The reciprocal of a number is 0.1. What is the number?** ....

5 **Express the following numbers in prime factor form:**

a) 36 ....   **b)** 64 ....   **c)** 930 ....

....

....

....

6 **a)** p is an even number. Sarah says that $\frac{1}{2}$ p + 1 always results in an even number. Give an example to show that Sarah is wrong ....

b) The letters q and r represent prime numbers. Give an example to show that q + r does not always result in an even number ....

Numbers 1 & 2 (cont)

7 a) What is the highest common factor of 20 and 36?

b) What is the least common multiple of 20 and 36?

8 a) What is the highest common factor of 64 and 100?

b) What is the least common multiple of 64 and 100?

9 Ron is organising a barbecue. Bread rolls are sold in packs of 30. Beefburgers are sold in packs of 16. He needs exactly the same number of bread rolls and beefburgers. What is the least number of packs he can buy for each?

10 a) The highest common factor of two numbers is 4. The least common multiple of the same two numbers is 60. What are the two numbers?

b) The highest common factor of three numbers is 15. The least common multiple of the same three numbers is 90. What are the three numbers?

11 Here are ten numbers: **9, 14, 25, 29, 41, 50, 61, 70, 84, 100**
a) Which three numbers are **i)** factors of 200, **ii)** multiples of 25, **iii)** multiples of 7, **iv)** prime numbers?
b) Which of the above numbers has the reciprocal 0.02?

12 What is the reciprocal of a) 100, **b)** $\frac{1}{100}$, **c)** 0.01, **d)** $\frac{99}{100}$?

13 Express the following numbers in prime factor form: a) 30, **b)** 100, **c)** 2 048

14 What is the highest common factor and least common multiple of...
a) 15 and 18, **b)** 40 and 58, **c)** 15, 18 and 24?

15 Kate, Tess and Pat are clapping out a rhythm. Kate claps every 3 beats. Tess claps every 5 beats. Pat claps every 9 beats. They all start by clapping at the same time. How many beats is it before they all clap at the same time again?

Numbers 3

1 **Solve the following without using a calculator. Show all your working.**

a)
```
   475
+   28
_____
```

b)
```
  23179
+  4830
_____
```

c)
```
   238
-   99
_____
```

d)
```
  43008
-  1559
_____
```

e)
```
   423
x   36
_____
```

f)
```
  4705
x   91
_____
```

g)
```
16 | 608
```

h)
```
13 | 3341
```

i) 41 x 100

j) 573 x 1000

k) 423 ÷ 10

l) 8 ÷ 1000

2 **a)** A whole number is multiplied by 1 000 giving an answer of 234 000

What is the number? ....

b) A whole number is divided by 100 giving an answer of 0.08.

What is the number? ....

3 **a)** The number 864 900 is divided by a power of 10 to give the answer 8.649.

What is the power of 10 used?

....

b) The number 763 is multiplied by a power of 10 to give the answer 76 300 000.

What is the power of 10 used?

....

4 **Solve the following without using a calculator. Show all your working.**
a) 3 001 + 999 b) 4 735 + 381 + 49 c) 4 079 - 497 d) 13 574 - 9 281 e) 473 x 67 f) 375 x 413
g) 4 037 ÷ 11 h) 47 748 ÷ 23 i) 940 x 10 j) 403 x 100 k) 11 x 1 000 l) 33 942 x 10 000
m) 408 ÷ 10 n) 55 ÷ 100 o) 6 ÷100 p) 33 942 ÷ 10 000

5 **A whole number is multiplied by a certain power of 10 giving an answer of 23 000. When the whole number is divided by the same power of 10 the answer is 2.3.** **a)** What is the whole number being used?
b) What is the power of 10 being used?

Integers 1 & 2

1 **a)** Put the following integers into ascending order (lowest to highest):

14, -3, -1, 5, 12, -7, -11, 2

....

b) Put the following integers into descending order (highest to lowest):

-230, 467, 165, -62, -162, 70, -320, 8

....

2 Complete the following boxes:

a) **i)** $1 - 3 + \boxed{} = 6$ **ii)** $6 + \boxed{} = 2$ **iii)** $-6 + \boxed{} = -11$

iv) $-4 - \boxed{} = -9$ **v)** $4 - \boxed{} = -4$ **vi)** $-3 - \boxed{} = 5$

vii) $-6 \times \boxed{} = 12$ **viii)** $5 \times \boxed{} = -20$ **ix)** $-6 \times \boxed{} = 6$

x) $-12 \div \boxed{} = -2$ **xi)** $36 \div \boxed{} = -9$ **xii)** $-45 \div \boxed{} = 9$

b) **i)** $-4 - 6 + \boxed{} = 2$ **ii)** $6 - \boxed{} - 3 = 11$ **iii)** $-8 - 3 - \boxed{} = 5$

iv) $\boxed{} + 2 - 5 = -9$ **v)** $14 - 1 + \boxed{} = -15$ **vi)** $4 - 11 + \boxed{} = -5$

c) **i)** $\dfrac{-6 + \boxed{}}{-3} = 7$ **ii)** $\dfrac{-6 + 2 - \boxed{}}{-5} = 4$ **iii)** $\dfrac{\boxed{} - 6 - 3}{-1} = 4$

iv) $\dfrac{-3 \times -5 \times \boxed{}}{-6} = 10$ **v)** $\dfrac{9 \times \boxed{} \times -2}{4 - 6} = 9$

d) **i)** $\boxed{} + \boxed{} = -4$ **ii)** $\boxed{} - \boxed{} = -3$ **iii)** $\boxed{} \times \boxed{} = -18$

iv) $\boxed{} \div \boxed{} = -10$ **v)** $\boxed{} \times -5 \times \boxed{} = 30$

3 Complete the following boxes by inserting =, −, x or ÷ to connect the numbers:

a) $6 \boxed{} 9 \boxed{} -3$

b) $2 \boxed{} 3 \boxed{} -7 \boxed{} 12$

c) $9 \boxed{} 1 \boxed{} -3 \boxed{} -3$

d) $-50 \boxed{} 10 \boxed{} 5 \boxed{} 0$

e) $6 \boxed{} -3 \boxed{} 10 \boxed{} 19 \boxed{} 1$

f) $10 \boxed{} -2 \boxed{} 6 \boxed{} 2 \boxed{} 0$

4 **The following table shows the highest temperatures (in °C) for eight places on a particular day:**

Place	Athens	Berlin	Cairo	Cardiff	London	Madrid	Moscow	New York
Temperature	15	-2	25	2	6	13	-6	18

a) What is the difference in temperature between the following places:

i) Athens and Berlin?...

ii) Cairo and Moscow? ....

iii) Berlin and Moscow?..

b) The next day the temperature in Cardiff had fallen by 2°C and the temperature in Moscow had fallen by 5°C. What was the difference between Cardiff and Moscow on that day?

....

5 **Below is part of a bank statement:**

LONSDALE BUILDING SOCIETY

Date	Description	Deposit	Withdrawal	Balance
11/12/03				£226.30
12/12/03	The Toy Shop		-£49.99	£176.31
13/12/03	Gas Bill		-£21.03	£155.28
13/12/03	Cheque	£25.00		£180.28
16/12/03	La Trattoria		-£32.98	☐
19/12/03	Rent		☐	-£112.70

a) What was the balance on 16/12/03 after the withdrawal of £32.98?

....

b) On 19/12/03 there was a withdrawal to pay for rent. How much was the withdrawal?

....

6 **Put the following integers into ascending order:**

3, -3, -11, 0, 4, 19, -36, 74, -1, 100

7 **Complete the following:**

a) 6 – 8 = **b)** -4 – 4 = **c)** -1 – 1 + 2 = **d)** -5 x -6 x 10 = **e)** -5 x 3 x 6 = **f)** -15 ÷ 3 x -4 = **g)** 100 ÷ 4 x -1 =

8 **Complete the following boxes:**

a) ☐ – 3 = 12 **b)** -8 – ☐ = 3 **c)** -6 + ☐ – 2 = 11 **d)** -14 + 20 + ☐ = -10 **e)** ☐ + 4 – 8 = -15

Order of Operations

1 **Calculate the following:**

a) 7 + 5 x 2

....

....

b) 14 x 5 – 3

....

....

c) 6^2 + 4 x 2

....

....

d) $\frac{15}{3}$ + 6

....

....

e) 18 x 2 – 5^2

....

....

f) $\frac{30}{6}$ – $\frac{40}{10}$

....

....

g) 2^3 + 4^3 x 3

....

....

h) 6 x 7 – 10 x 3 + 8

....

....

i) 15 – 3 x 6 x 2 + 5

....

....

2 **a)** Put brackets in the following expression so that its value is 43:

13 – 3 x 4 + 3

b) Put brackets in the following expression so that its value is -8:

13 – 3 x 4 + 3

c) Put brackets in the following expression so that its value is 70:

13 – 3 x 4 + 3

d) Put brackets in the following expression so that its value is 4:

13 – 3 x 4 + 3

e) Put brackets in the following expression so that its value is 20.5:

3 x 1.4 + 4 x 2.5

f) Put brackets in the following expression so that its value is 12.12:

7 + 3.2 x 6 – 4.4

---

3 **Calculate the following:**
   a) 4 + 3 x 13   **b)** 4 x 3 – 13   **c)** 4 – 3 x 13   **d)** 4^2 ÷ 2 + 3   **e)** 9 + 15 ÷ 5   **f)** 4 + 7 x 3 – 15
   g) 14 ÷ 4 – 5   **h)** 3 + 7 – 5 x 3^2

4 **Put brackets in the following expression so that its value is... a)** -1   **b)** 35   **c)** -51

   3^2 – 4 x 5 + 10

5 **Put brackets in the following expression so that its value is... a)** 54   **b)** -2   **c)** 6

   4^2 - 3 x 4 + 2

Rounding Numbers 1 & 2

1 **Round ...**

a) 7.321 to **i)** 1 decimal place .... **ii)** 2 decimal places ....

b) 16.781 to **i)** 1 decimal place .... **ii)** 2 decimal places ....

c) 0.01765 to **i)** 2 decimal places .... **ii)** 3 decimal places ....

d) 0.1053 to **i)** 1 decimal place .... **ii)** 3 decimal places ....

e) 7.0707 to **i)** 1 decimal place .... **ii)** 2 decimal places ....

2 **Dave weighs 68.4kg to 1 decimal place. What is the lowest measurement and the highest measurement possible for his actual weight to 2 decimal places?**

Lowest possible weight .... Highest possible weight ....

3 **Helen's height is 1.65m to 2 decimal places. What is the lowest measurement and the highest measurement possible for her actual height to 3 decimal places?**

Lowest possible height .... Highest possible height ....

4 **Round...**

a) 432 to **i)** 1 significant figure .... **ii)** 2 significant figures ....

b) 9 154 to **i)** 1 significant figure .... **ii)** 2 significant figures ....

c) 10 047 to **i)** 2 significant figures .... **ii)** 4 significant figures ....

d) 0.0238 to **i)** 1 significant figure .... **ii)** 2 significant figures ....

e) 0.0001736 to **i)** 2 significant figures .... **ii)** 3 significant figures ....

f) 0.010366 to **i)** 2 significant figures .... **ii)** 3 significant figures ....

5 **The attendance at a pop concert was 5 700 to 2 significant figures. What is the difference between the lowest possible actual attendance and the highest possible actual attendance?**

....

....

6 **Round 135.6742 to a)** 1 decimal place **b)** 2 decimal places **c)** 1 significant figure
d) 2 significant figures **e)** 3 significant figures

7 **Round 0.06038 to a)** 1 decimal place **b)** 2 decimal places **c)** 3 decimal places
d) 1 significant figure **e)** 2 significant figures **f)** 3 significant figures

8 **Fran's weight to 1 decimal place is 47.6kg. What is the difference between her highest possible weight and her lowest possible weight to 2 decimal places?**

9 **a)** Use your calculator to work out the value of $(1.46)^2$ x 6.71. Write down all the digits on your display.
b) Round your answer to a suitable degree of accuracy.

Estimating & Checking

1 **a) i)** Work out 106 x 53 using a calculator ....

ii) Without using a calculator check your answer by estimation.

....

b) i) Work out 3.8 x 15.2 using a calculator ....

ii) Without using a calculator check your answer by estimation.

....

c) i) Work out $\frac{29.4 \times 5.8}{19.6}$ using a calculator ....

ii) Without using a calculator check your answer by estimation.

....

d) i) Work out 384 726 x 0.00071 using a calculator ....

ii) Without using a calculator check your answer by estimation.

....

2 **John and Donna collect and keep all their loose change. The table below shows how much they collected for three successive months.**

MONTH	JOHN	DONNA
JUNE	£7.36	£9.10
JULY	£8.90	£16.58
AUGUST	£13.47	£4.52

a) John calculated that he collected £29.73 altogether.

i) Check by estimation whether this is likely to be correct.

....

....

ii) Check John's calculation for accuracy.

....

....

b) Donna calculated that she collected £31.20 altogether.

i) Check by estimation whether she is likely to be correct.

....

ii) Check Donna's calculation for accuracy.

....

....

---

3 **Check the following calculations by estimation and then for accuracy. Finish by making any necessary corrections**
a) 13 + 29 + 43 = 82 **b)** 9.06 + 11.58 + 7.23 + 13.86 = 41.73 **c)** (3 x 6.25) + 92 − 10 = 101.75
d) (4.7 x 6.3) + (0.8 x 9.5) = 40.21

4 **a)** Calculate $\frac{8.7}{3.2 - 1.9}$ **b)** Give your answer to an appropriate degree of accuracy.
c) Without using a calculator, check your answer by estimation.

1 **Work out the value of...**

   a) 2^3 ....   **b)** 3^2 ....   **c)** 4^3 ....

   d) 10^3 ....   **e)** 3^4 ....   **f)** 1^5 ....

2 **Here are ten numbers:**

$$10 \quad 18 \quad 25 \quad 27 \quad 45 \quad 64 \quad 80 \quad 125 \quad 133 \quad 196$$

   a) Which three numbers are square numbers? ....

   b) Which three numbers are cube numbers? ....

   c) Which one of these numbers is a square number and a cube number? ....

   d) Which of the above numbers is equal to $10^2 - 6^2$? ....

   e) Which of the above numbers is equal to $4^3 + 2^4$? ....

3 **Work out the value of...**

   a) $(-3)^2$ ....   **b)** $(-3)^3$ ....   **c)** $(-5)^2$ ....

   d) $(-5)^3$ ....   **e)** $(-1)^2$ ....   **f)** $(-1)^3$ ....

4 **Find the value of ...**

   a) $2^3 \times 2^2$ ....   **b)** $3^2 \times 3^1$ ....

   c) $4^3 \times 4^2 \times 4^1$ ....   **d)** $2^3 \div 2^2$ ....

   e) $10^4 \div 10^2$ ....   **f)** $6^4 \div 6^0$ ....

   g) $(2^2)^2$ ....   **h)** $(4^3)^2$ ....

   i) $(10^2)^3$ ....   **j)** $(2^2 \times 2^3)^2$ ....

   k) $\dfrac{10^2 - 6^2}{4^3}$ ....   **l)** $\dfrac{4^3 + 2^6}{8^2}$ ....

   m) $\dfrac{(4^1)^3}{8^2}$ ....   **n)** $\dfrac{2^2 \times 2^1 \times 2^4}{4^3}$ ....

5

   a) What is $3^2 \times 9$ as a single power of 3? ....

   b) What is $4^2 \times 2^3$ as a single power of 2? ....

   c) What is $8^2 \times 2^4$ as a single power of 4? ....

6 **Work out the value of...**

   a) 4^4 **b)** 4^1 **c)** $4^4 \times 4^1$ **d)** $4^3 \times 5^2$ **e)** $8^0 + 3^3$ **f)** $6^2 - 3^2$ **g)** $\dfrac{8^2}{2^5}$ **h)** $\dfrac{3^3 - 7^1}{2^2}$ **i)** $\dfrac{10^2 + 3^3 + 5^0}{8^2}$

7 **a)** What is $3^3 \times 9^2$ as a single power of 3?   **8** **a)** Work out the cube of 5.   **b)** Work out $5 \div 0.9^2$ **(i)** Write down the full

   b) What is $5^2 - 3^2$ as a single power of 2?   calculator display.   **(ii)** Give your answer to the nearest whole number.

Roots

1 **Work out the value of...**

a) $\sqrt{25}$ ....

b) $36^{\frac{1}{2}}$ ....

c) $\sqrt{144}$ ....

d) $196^{\frac{1}{2}}$ ....

e) $64^{\frac{1}{3}}$ ....

f) $\sqrt[3]{1}$ ....

g) $8^{\frac{1}{3}}$ ....

h) $\sqrt[3]{1000}$ ....

i) $27^{\frac{1}{3}}$ ....

2 **Here are ten numbers:**

$$2 \qquad 4 \qquad 5 \qquad 8 \qquad 10 \qquad 20 \qquad 27 \qquad 36 \qquad 64 \qquad 80$$

a) Which three numbers have an integer square root? ....

b) Which three numbers have an integer cube root? ....

c) Which one of these numbers has an integer square root and an integer cube root? ....

d) Which of these numbers is equal to $\sqrt{144} - 8^{\frac{1}{3}}$? ....

e) Which of these numbers is equal to $81^{\frac{1}{2}} \times \sqrt{16}$? ....

3 **Work out the value of...**

a) $(\sqrt{4})^2$ ....

b) $(\sqrt[3]{8})^2$ ....

c) $\sqrt{4^2 \times 25}$ ....

d) $100^{\frac{1}{2}} \times \sqrt[3]{27}$ ....

e) $64^{\frac{1}{2}} \times \sqrt[3]{64}$ ....

f) $27^{\frac{2}{3}}$ ....

g) $16^{\frac{3}{2}}$ ....

4 a) What is $\sqrt{4} \times 8$ as a single power of 4? ....

b) What is $\sqrt[3]{27} \times 3$ as a single power of 3? ....

c) What is $\sqrt[3]{64} \times 16^{\frac{1}{2}}$ as a single power of 2? ....

d) What is $\sqrt{9} \times 1^3 \times \sqrt{144}$ as a single power of 6? ....

5 **Simplify the following surds:**

a) $\sqrt{27}$ ....

b) $\sqrt{50}$ ....

c) $\sqrt{54}$ ....

d) $\sqrt{90}$ ....

---

6 **Work out the value of ...**

a) $169^{\frac{1}{2}}$ b) $\sqrt[3]{125}$ c) $\sqrt{81}$ d) $125^{\frac{1}{3}}$ e) $225^{\frac{1}{2}}$

7 **Find the value of ...**

a) $(\sqrt{9})^3$ b) $(\sqrt[3]{8})^2$ c) $\sqrt[3]{6^2 + 28}$ d) $4^{\frac{3}{2}}$ e) $64^{\frac{2}{3}}$

8 **Simplify the following surds:**

a) $\sqrt{200}$ b) $\sqrt{80}$ c) $\sqrt{63}$

Standard Form

1 Write the following examples of standard form as ordinary numbers:

a) 2.3×10^2 ....

b) 2.3×10^3 ....

c) 4.21×10^4 ....

d) 6.32×10^1 ....

e) 7.467×10^3 ....

f) 3×10^{-2} ....

g) 2.3×10^{-3} ....

h) 4.21×10^{-4} ....

i) 6.324×10^{-2} ....

2 Write the following numbers in standard form:

a) 600 ....

b) 473 ....

c) 42 000 ....

d) 413 256 ....

e) 496.3 ....

f) 0.032 ....

g) 0.47 ....

h) 0.000631 ....

i) 0.1 ....

3 A bricklayer has 9 800kg of sand.

a) Write 9 800kg in grams. Give your answer in standard form.

....

b) One grain of sand weighs 0.00004g. Write this weight in standard form.

....

....

c) How many grains of sand are there in 9 800kg of sand? Give your answer in standard form.

....

....

4 Calculate the following. Give your answers in standard form.

a) $9.23 \times 10^2 + 4.71 \times 10^3$ ....

b) $7.15 \times 10^5 - 9.68 \times 10^4$ ....

c) $2.34 \times 10^4 \times 3.6 \times 10^7$ ....

d) $5.5 \times 10^7 \div 1.1 \times 10^9$ ....

5 In 1901 the population of England and Wales was 3.26×10^7. If the area of England and Wales is 151 000km^2 (to the nearest thousand), calculate what the population per square kilometre was in 1901. Give your answer to 2 significant figures.

6 In one minute light will travel a distance of approximately 1.8×10^{10} metres.
a) How far will light travel in 1 hour? Give your answer in standard form. **b)** How far will light travel in 1 year? Give your answer in standard form. **c)** The sun is approximately 1.44×10^{11} metres away. How long does it take light to travel from the sun to the earth? **d)** A light year is the distance travelled by light in one earth year. Our nearest star, after the sun, is 4.3 light years away. How far away is this star in metres? Give your answer in standard form.

7 Here are five numbers written in standard form. 3.2×10^6 1.05×10^7 4.91×10^0 9.6×10^{-3} 5.2×10^{-1}
a) Write down **(i)** the largest number   **ii)** the smallest number.   **b)** Write down 9.6×10^{-3} as an ordinary number.
c) Work out $3.2 \times 10^6 \div 0.1$. Give your answer in standard form.

Fractions 1 & 2

1 **Here are eight fractions:**

$$\frac{35}{50}, \frac{16}{40}, \frac{60}{90}, \frac{28}{40}, \frac{30}{40}, \frac{40}{100}, \frac{84}{120}, \frac{10}{25}$$

a) Which three fractions are equivalent to $\frac{2}{5}$? ....

b) Which three fractions are equivalent to $\frac{7}{10}$? ....

2 **Express the following fractions in their simplest form:**

a) $\frac{27}{30}$ ....
b) $\frac{42}{6}$ ....
c) $\frac{84}{105}$ ....
d) $\frac{108}{184}$ ....

3 a) Arrange the following fractions in ascending (lowest to highest) order:

$$\frac{5}{6}, \quad \frac{3}{5}, \quad \frac{11}{15}, \quad \frac{2}{3}, \quad \frac{1}{2}$$

....

b) Arrange the following fractions in descending (highest to lowest) order:

$$\frac{9}{40}, \quad \frac{3}{5}, \quad \frac{5}{8}, \quad \frac{9}{10}, \quad \frac{1}{4}$$

....

4 **Write down two fractions that are greater than $\frac{7}{10}$ but less than $\frac{5}{6}$.**

....

5 a) Write the following improper fractions as mixed numbers:

i) $\frac{11}{5}$ ....
ii) $\frac{13}{6}$ ....
iii) $\frac{24}{5}$ ....
iv) $\frac{32}{3}$ ....

b) Write the following mixed numbers as improper fractions:

i) $3\frac{1}{3}$ ....
ii) $5\frac{1}{4}$ ....
iii) $11\frac{3}{5}$ ....
iv) $20\frac{1}{20}$ ....

6 Write down three fractions that are equivalent to each of the following:

a) $\frac{2}{3}$ b) $\frac{4}{7}$ c) $\frac{9}{11}$

7 Express the following fractions in their simplest form and then arrange them into ascending order:

$$\frac{28}{35}, \quad \frac{38}{40}, \quad \frac{75}{100}, \quad \frac{42}{84}, \quad \frac{99}{110}$$

8 Write down three fractions that are greater than $\frac{4}{5}$ but less than $\frac{9}{10}$.

Fractions 3

1 Solve the following additions and subtractions without using a calculator. Show all your working and give your answers in their simplest form.

a) $\frac{3}{4} + \frac{2}{3}$

b) $\frac{2}{9} + \frac{7}{8}$

c) $4\frac{1}{2} + 2\frac{9}{10}$

d) $7\frac{5}{8} + 4\frac{1}{3}$

e) $\frac{9}{10} - \frac{1}{2}$

f) $\frac{13}{15} - \frac{2}{3}$

g) $4\frac{4}{5} - 1\frac{3}{8}$

h) $9\frac{1}{6} - 4\frac{3}{5}$

2 At a football match the crowd is made up as follows:

$\frac{5}{12}$ of the crowd are over 40 years old, $\frac{1}{4}$ of the crowd are between 20 years old and 40 years old and the remainder of the crowd are less than 20 years old.

What fraction of the crowd are less than 20 years old? Give your answer in its simplest form.

....

....

....

3 Solve the following multiplications and divisions without using a calculator. Show all your working again and give your answers in their simplest form.

a) $\frac{1}{4} \times \frac{2}{5}$

b) $\frac{9}{10} \times \frac{2}{7}$

c) $\frac{3}{4} \div \frac{9}{10}$

d) $\frac{7}{8} \div \frac{7}{12}$

e) $3\frac{1}{2} \times 1\frac{3}{5}$

f) $3\frac{1}{4} \times 1\frac{5}{7}$

g) $4\frac{1}{2} \div 2\frac{2}{3}$

h) $1\frac{1}{2} \div 2\frac{1}{4}$

i) $\frac{8}{9} \times 4$

j) $\frac{7}{10} \div 3$

4 Cassie thinks $\frac{1}{5} + \frac{1}{3} = \frac{2}{8}$ She is wrong. Show the correct way to work out $\frac{1}{5} + \frac{1}{3}$

5 Solve the following without using a calculator. Show all your working.

a) $\frac{7}{10} + \frac{5}{8}$ b) $2\frac{1}{3} + 4\frac{3}{8}$ c) $\frac{9}{10} - \frac{2}{3}$ d) $4\frac{1}{2} - 3\frac{7}{10}$ e) $\frac{2}{5} \times \frac{7}{10}$ f) $\frac{13}{15} \div \frac{4}{5}$

g) $3\frac{1}{4} \times 1\frac{2}{3}$ h) $4\frac{1}{5} \div 3\frac{1}{2}$ i) $4\frac{1}{2} \times 5$ j) $2\frac{1}{3} \div 6$

Lonsdale REVISION GUIDES   Revision Guide Reference: Page 20   NUMBER   17

Calculations Involving Fractions

1 James wants to buy a CD player costing £240.

As a deposit he pays $\frac{1}{8}$ of the cost. How much deposit did James pay?

....

....

2 Mary buys a new car costing £13 000.

As a deposit she pays $\frac{2}{5}$ of the cost. The remainder she pays monthly over a period of 4 years.

a) How much deposit does Mary pay?

....

....

b) What is her monthly repayment?

....

....

c) At the end of the 4 years Mary decides to sell her car. It is now worth £7 150. Express its value now, after 4 years, as a fraction of its value when new. Give your answer in the simplest form.

....

....

3 A shop holds a sale where all items are reduced by $\frac{1}{6}$

a) Calculate the sale price of an item which cost £93.60 before the sale.

....

....

....

b) The sale price of another item is £121.50. Calculate its price before it was reduced.

....

....

4 A football match lasts $1\frac{1}{2}$ hours. During the match the ball is out of play for a total of 15 minutes. Express the total length of time that the ball is in play as a fraction of the time that the match lasts. Give your answer in its simplest form.

5 In a sale, a washing machine has its original price reduced by $\frac{1}{2}$. The following week the sale price is further reduced by $\frac{1}{4}$.

a) If the washing machine originally cost £600, calculate its sale price after **i)** the first reduction, **ii)** the second reduction.
b) If the washing machine costs £150 after both reductions what was its original price?

6 Express 40 seconds as a fraction of 1 hour.

1 Complete the following table (the first row has been done for you).

	Place Value of Digits							Decimal Number
	100 Hundreds	**10** Tens	**1** Units	DECIMAL POINT	$\frac{1}{10}$ Tenths	$\frac{1}{100}$ Hundredths	$\frac{1}{1000}$ Thousandths	
			3	•	4	2		3.42
a)	1	0	2		5			
b)		1	3		4	7	1	
c)								8.407
d)		9	0		0	3	1	
e)								423.008

2 Use a calculator to write the following fractions as either recurring or terminating decimals.

If the decimal is recurring, place a dot (·) over the digit or digits that repeat continuously.

a) $\frac{2}{3}$ .... b) $\frac{2}{5}$ .... c) $\frac{1}{11}$ .... d) $\frac{7}{9}$ ....

3 Arrange the following decimals in ascending order of value:

$$6.3, \ 0.36, \ 3.6, \ 0.306, \ 0.63$$

....

4 Complete the following additions and subtractions without using a calculator.

Show all your working.

a) 13.62 + 7.77 b) 103.2 + 4.837 c) 40.75 − 8.29 d) 723.4 − 6.19

5 Write the following fractions as decimals. For each recurring decimal, place a dot (·) over the digit or digits that repeat continuously.

a) $\frac{3}{8}$ b) $\frac{2}{9}$ c) $\frac{1}{30}$ d) $\frac{22}{25}$ e) $\frac{4}{15}$

6 Arrange the following decimals in descending (highest to lowest) order:
14.32, 1.432, 143.2, 13.42, 14.23, 1.342

7 Mrs Green goes shopping. She buys four tins of baked beans at 37p each, 3 tins of spaghetti at 29p each and 2 boxes of cornflakes at £1.37 each. She pays for her goods with a £10 note. How much change does she receive?

8 Peter puts three boxes in the boot of his car. They weigh 8.4kg, 6.73kg and 13.03kg. What is the total weight of the load?

Multiplication of Decimals

1 Solve the following without using a calculator. Where possible show all your working.

a) 4.7 x 10 **b)** 13.246 x 10 **c)** 0.00146 x 100 **d)** 136.3 x 1 000

....

e) 7.56 x 13 **f)** 4.72 x 2.3

g) 16.56 x 17.3 **h)** 4.713 x 1.56

2 If **27 x 36 = 972** write down, without making any further calculations, the value of...

a) 2.7 x 36 .... **b)** 27 x 0.36 .... **c)** 2.7 x 3.6 ....

d) 0.027 x 36 .... **e)** 0.27 x 0.36 ....

3 If **231 x 847 = 195 657** write down, without making any further calculations, the value of...

a) 0.231 x 847 .... **b)** 231 x 84 700 ....

c) 0.231 x 847 000 .... **d)** 2 310 x 8.47 ....

4 Solve the following without using a calculator.
Where possible show all your working.
a) 15.67 x 10 **b)** 0.0101 x 100 **c)** 3.4671 x 1000 **d)** 2.32 x 11
e) 4.67 x 1.8 **f)** 146.2 x 2.45 **g)** 13.33 x 0.23 **h)** 9.4 x 0.003

5 If **3.52 x 4.7 = 16.544** write down, without making any further
calculations, the value of...
a) 352 x 47 **b)** 3.52 x 47 **c)** 0.352 x 4.7 **d)** 3.52 x 0.047

6 A school holds a raffle. The three prizes cost £24.65, £17.99 and
£9.89. 130 tickets were sold at £1.25 each.
How much profit did the school make from the raffle?

7 Jim is going to hire a cement mixer. The cost is £24.50 for the
first day and £3.75 for each extra day. Jim wants to hire it for 7
days. How much will it cost him in total?

8 A shirt costs £24.95. How much is this in euros if £1 = €1.6?

9 Books are stored in piles of 200. Each book is 0.5cm thick.
Calculate the height of one pile of books without using a
calculator.

£24.95

Division of Decimals

1 Solve the following without using a calculator. Where possible show all your working.

   a) 16.3 ÷ 10   **b)** 0.347 ÷ 10   **c)** 14 632.4 ÷ 100   **d)** 1.2467 ÷ 1 000

   ....

   e) 37.6 ÷ 8   **f)** 41.04 ÷ 1.2

   g) 1 141.72 ÷ 0.23   **h)** 549.6 ÷ 1.2

2 If 72.8 ÷ 56 = 1.3 write down, without making any further calculations, the value of...

   a) 72.8 ÷ 5.6 ....   **b)** 7.28 ÷ 56 ....   **c)** 72.8 ÷ 0.56 ....

   d) 0.728 ÷ 56 ....   **e)** 728 ÷ 56 ....

3 If 27 x 34 = 918 write down, without making any further calculations, the value of...

   a) 918 ÷ 27 ....   **b)** 918 ÷ 3.4 ....   **c)** 91.8 ÷ 27 ....

   d) 9.18 ÷ 3.4 ....   **e)** 91.8 ÷ 340 ....

---

4 Solve the following without using a calculator. Where possible show all your working.
   a) 7.162 ÷ 10   **b)** 0.0034 ÷ 100   **c)** 473.1 ÷ 1 000   **d)** 707.4 ÷ 9   **e)** 24.64 ÷ 1.4   **f)** 0.08788 ÷ 0.0013

5 If 58.82 ÷ 3.4 = 17.3 write down, without making any further calculations, the value of...
   a) 58.82 ÷ 0.34   **b)** 5 882 ÷ 0.34   **c)** 58.82 ÷ 17.3   **d)** 3.4 x 17.3

6 A school holds a raffle. The three prizes cost £9.49, £14.99 and £19.99. Tickets cost 75p each.
   What is the minimum number of tickets that need to be sold for the school to make a profit?

7 Jean is going to hire a wallpaper stripper. The cost is £8.50 for the first day and £1.25 for each extra day.
   When she returns the wallpaper stripper the total hire charge is £24.75. For how many days did she hire the
   wallpaper stripper?

Percentages 1

1 Calculate the following amounts:

    a) 20% of 60p

    b) 30% of 6.5km

....

....

2 Express ...

    a) £18 as a percentage of £90

    b) 42cm as a percentage of 8.4m

....

....

3 Alan has kept a record of his height and weight from when he was age 10 and age 16.

Age 10	1.2m tall	40kg weight
Age 16	1.74m tall	64kg weight

    a) i) Calculate the increase in his height from age 10 to age 16 ....

    ii) Express this increase as a percentage of his height at age 10.

....

....

    b) i) Calculate the increase in his weight from age 10 to age 16 ....

    ii) Express this increase as a percentage of his weight at age 10.

....

....

4 A train ticket costs £17.60 when bought on the day of travel. If the same ticket is bought in advance it costs £15.40. Express the saving you make when you buy the ticket in advance as a percentage of the full ticket price when bought on the day of travel.

....

....

....

5 Calculate the following amounts:
    a) 45% of £2.60, **b)** 80% of 6.4kg, **c)** 5% of £10.40.

6 **a)** Express 46cm as a percentage of 69cm, **b)** Express 200m as a percentage of 200km,
    c) Express 550g as a percentage of 2kg.

7 A tin of tomato soup weighs 420g. Special tins weigh 546g. Calculate the increase in the weight of a special tin as a percentage of the weight of a normal tin.

8 A brand new car bought in the UK costs £18 000. The same car bought abroad costs £14 850. Calculate the decrease in the price when the car is bought abroad as a percentage of its price in the UK.

Percentages 2

1 A brand new car costs £15 000. It is estimated that it will depreciate in value by 23% in the first year. What is the value of the car at the end of the first year?

....

....

2 Mr Dixon earns £320 per week. He is given a 3% pay rise. Mrs Asaf earns £340 per week. She is given a 2.5% pay rise. Whose weekly pay increases by the greatest amount? You must show all your working.

....

....

....

3 Mr and Mrs Smith bought a house for £120 000. Each year since, the value of the property has appreciated by 10%.

a) Calculate the value of the house 2 years after they bought it.

....

....

....

b) The value of the house after two years can be found by multiplying the original cost by a single number. What is this single number as a decimal?

....

4 Jim buys a new motorbike for £10 000. **a)** If the value of the motorbike decreases by 20% each year, calculate the value of the motorbike after three years.

....

....

....

....

b) The value of Jim's bike after three years can be found by multiplying the original cost by a single number. What is this single number as a decimal?

....

5 An electrical shop has a sale. All items are reduced by 20%. The following week the shop takes 10% off its sale prices. Pat wants to buy a fridge that was priced at £100 before the sale. She reckons that she will save 30% of this price and that the fridge will now cost her £70. Is she correct? Explain why.

6 Bob weighs 120kg on January 1. Over the first six months of the year his weight increases by 5%. Over the next six months his weight decreases by 10%.
a) What is his weight at the end of the year?
b) Bob's weight at the end of the year can be found by multiplying his weight on January 1 by a single number. What is this single number as a decimal?

Percentages 3

1 Phil has been told by his mum that he needs to spend 1 hour a day doing his homework, an increase of 50%. How long did Phil originally spend doing his homework each day?

....

....

....

2 An electrical shop has a sale. All items are reduced by 15%. A tumble drier has a sale price of £122.40. What was the price of the tumble drier before the sale?

....

....

....

3 A man buys an antique clock. He later sells it for £5 040, an increase of 12% on the price he paid for it. How much did the clock cost him?

....

....

....

....

4 Dave is a long distance lorry driver. On Tuesday he drives 253km. This is a 15% increase on the distance he drove the previous day. How far did he drive on Monday?

....

....

....

5 Jean has her house valued. It is worth £84 000. This is a 40% increase on the price she originally paid for it. How much did Jean pay for her house?

....

....

....

6 A clothes shop has a sale. All prices are reduced by 30%. The sale price of a dress is £86.80. What was the original price of the dress?

7 Mrs Smith buys some shares. In twelve months their value has increased by 15% to £3 680. How much did she pay for the shares?

8 Mr Jones collects stamps. After two years he has increased the number of stamps in his collection by 120% to 660. How many stamps did he have in his collection two years ago?

Converting Between Systems

1 Complete the following table. The first row has been done for you.

	Fraction (simplest form)	Decimal	Percentage
	$\frac{1}{2}$	0.5	50%
a)	$\frac{3}{10}$		
b)		0.45	
c)			37.5%
d)	$\frac{2}{3}$		
e)		0.125	
f)			84%
g)	$1\frac{4}{5}$		
h)		4.6	
i)			225%

2 $\frac{3}{5}$ of the CDs in Peter's collection are pop music. Is this more or less than 55% of his collection? Explain your answer.

....

....

3 Write these numbers in ascending order.   44%, $\frac{2}{5}$,   35%,   0.42, $\frac{1}{3}$,   0.25

....

....

4 Which of these is the largest amount? You must show all working.   70% of £50, $\frac{4}{10}$ of £90

....

....

5 a) Write the following as decimals: **i)** $\frac{4}{5}$ **ii)** 90%   **iii)** $\frac{17}{20}$ **iv)** 85%

b) Write these numbers in descending order: $\frac{19}{20}$, 0.75, 90%, 0.92, $\frac{4}{5}$, 85%

6 a) Write the following as percentages: **i)** 0.62 **ii)** $\frac{3}{5}$ **iii)** 0.56 **iv)** $\frac{29}{50}$

b) Write the following numbers in ascending order: 61%, $\frac{3}{5}$, 0.56, 0.62, $\frac{29}{50}$, 65%

Everyday Maths 1 & 2

1 **a)** Alma buys a washing machine. The price is £380 + VAT at 17.5%.

How much does the washing machine cost in total?

....

....

b) The total cost of a home stereo system is £517 including VAT at 17.5%.

What is the price of the home stereo system before VAT?

....

....

2 **Mr Smith wants to invest £20 000 for two years. He has two options:**

SIMPLE INTEREST AT 5% OR COMPOUND INTEREST AT 4.8%

a) Which option will make him the most money? Show all your working.

....

....

....

b) Which option will make him the most money after five years?
Show all your working.

....

....

....

....

....

....

....

3 **Peter wants to buy a car. There are two payment options:**

Cash price £8 000 or 30% deposit + 24 monthly payments of £255

a) What is the deposit required for hire purchase?

....

b) Calculate the percentage increase if Peter pays for the car by hire purchase compared to buying the

car at the cash price.

....

....

....

4 **Mr Spark receives an electricity bill. Complete the bill by filling in the gaps.**

Meter Reading				
Present	Previous	Units Used	Pence per unit	Amount (£)
32467	30277	i)	8p	ii) ....
			Quarterly charge	9.60
		Total charged this quarter excluding VAT		iii) ....
		VAT at 5%		iv) ....
		Total payable		v) ....

5 **Use the timetable alongside to answer the following questions:**

a) Grace wants to arrive in London before midday. What is the latest train she can catch from Millford station to get there in time and how long will her train journey take?

....

....

b) Simon needs to get to Woking by 11.15am because he has a job interview. What is the time of the last train he can catch from Farncombe?

....

c) What percentage of trains departing from Woking station take less than 30 minutes to arrive at London Waterloo?

....

....

Petersfield, Millford, Farncombe, Woking to London Waterloo

Mondays to Fridays

	AN	NW	AN	AN	AN
Petersfield	0752	0811	0833	0901	0928
Liphook	—	—	—	—	—
Haslemere	—	—	—	—	—
Witley	—	—	—	—	—
Millford (Surrey)	0806	0829	0845	0917	0941
Godalming	—	—	—	—	—
Farncombe	0822	0850	0900	0937	0959
Guildford	—	—	—	—	—
Reading	—	—	—	—	—
Woking	0830	0900	0907	0947	1007
Heathrow Airport (T1)	—	—	—	—	—
Clapham Junction	—	—	—	—	—
London Waterloo	0903	0932	0939	1018	1036

Mondays to Fridays

	AN	NW	AN	AN	AN
Petersfield	0949	0956	1019	1049	1055
Liphook	—	—	—	—	—
Haslemere	—	—	—	—	—
Witley	—	—	—	—	—
Millford (Surrey)	1002	1011	1032	1102	1114
Godalming	—	—	—	—	—
Farncombe	1017	1032	1047	1117	1132
Guildford	—	—	—	—	—
Reading	—	—	—	—	—
Woking	1026	1043	1059	1128	1142
Heathrow Airport (T1)	—	—	—	—	—
Clapham Junction	—	—	—	—	—
London Waterloo	1052	1111	1125	1155	1211

6 What is the total cost of a vacuum cleaner if the price is £180 + VAT at 17.5%?

7 Mr Walker wants to invest £20 000 for 3 years. He can either invest it at 6% simple interest or 5.5% compound interest. Which option will make him the most money? Show all your working.

8 Mr Brum wants to buy a car costing £6 000. He buys it on hire purchase paying £124 a month for 3 years. His total repayment is £6 264. What deposit did he pay as a percentage of the car's value?

9 Mr Plug receives an electricity bill. The cost per unit is 8p and the quarterly charge is £9.60. It says on his bill that the total payable for this quarter excluding VAT is £80.24. How many units has he used this quarter?

10 Using the timetable above, what is the difference between the average journey time of Peak time trains and the average journey time of Off Peak trains travelling from Petersfield to London Waterloo? Trains that arrive at London Waterloo after 10am are classified as Off Peak.

Ratio & Proportion 1 & 2

1 Jim weighs 70kg. His sister Cathy weighs 35kg.

a) Calculate the ratio of Jim's weight to Cathy's weight.

....

b) Express your answer to part a) in the form 1:n ....

2 Freya is 90cm tall. Her brother Tom is 1.35m tall.

a) Calculate the ratio of Freya's height to Tom's height.

....

b) Express your answer to part a) in the form 1:n ....

3 $\frac{3}{7}$ of the teachers at a school are male. a) What is the ratio of male teachers to female teachers?

....

b) Express your answer to part a) in the form 1:n ....

4 A large tin of baked beans costs 36p and weighs 450g. A small tin of baked beans costs 22p and weighs 250g.

a) Calculate the ratio of the weight of the two tins ....

b) Calculate the ratio of the cost of the two tins...

c) Which tin represents the best value for money? Explain your choice.

....

....

5 a) Divide 80p in the ratio of 2:3   b) Divide 6m 30cm in the ratio 2:3:4

....

....

....

....

....

6 £5 000 is shared between three women in the ratio of their ages. Their combined age is 120 years. If Susan gets £2 500, Janet gets £1 500 and Polly gets the remainder, what are their ages?

....

....

....

....

....

Ratio & Proportion 1 & 2 (cont)

7 In a maths class there are 30 pupils on the register and the ratio of girls to boys is 3:2. If 4 girls and 2 boys are absent from the class what does the ratio of girls to boys become?

....
....
....
....
....
....

8 A builder makes concrete by mixing cement, gravel, sand and water in the ratio 2:8:5:3 by weight. How many kilograms of sand, to the nearest kg, does he need to make 10 000kg of concrete?

....
....
....
....
....
....

9 Opposite is a recipe for making 10 biscuits.

a) Calculate the amount of each ingredient needed to make 25 biscuits.

....
....
....
....

b) How many biscuits can be made using 0.715kg of oatmeal?

....
....
....
....

Recipe
90g of flour
130g oatmeal
80g margarine

---

10 100ml of semi-skimmed milk contains 4.8g carbohydrates and 1.8g fat.
a) What is the ratio of carbohydrates to fat? **b)** Express your answer to part a) in the form 1: n.

11 A large tub of margarine weighs 500g and costs £1.10. A small tub of margarine which normally weighs 250g has an extra 10% free and costs 52p. Which tub of margarine represents the best value for money? Explain.

12 Mr Thorpe inherits £15 000. He divides the money between his four children, Lucy, Paul, John and Sarah in the ratio 6: 7: 8: 9 respectively. How much do they each receive?

13 450 tickets were sold for a raffle at 20p each. The ratio of the cost of prizes to profit made is 5:13. How much profit did the raffle make?

14 The angles of a quadrilateral are in the ratio 2:3:5:8. What is the size of the largest angle?

15 A large packet of washing powder weighs 2.5kg and costs £5.60. How much should a 750g packet of washing powder cost if it represents the same value as the large packet?

The Basics of Algebra 1 & 2

1 Simplify…

a) $a + 2a$

b) $5x + 6x$

c) $11a - 6a$

d) $9p - 2p + 4p$

e) $2a^2 + 3a^2$

f) $10w^2 - w^2 + 2w^2$

g) $2a + 3b + 4a + 5b$

h) $7c - 8d - 9c + 10d$

i) $3a \times 2b$

j) $12pq \times 2r$

k) $4ab + 8cd - 7ab + cd + ab$

l) $-4x^2 + 6x + 7 + 3x^2 - 11x + 2$

m) $14 - 8p^2 + 11p + 2p^2 - 4p + 3$

2 **Explain why $2x + 3x = 5x$ is an example of an identity and not an equation.**

3 Simplify…

a) $a^2 \times a^3$

b) $4b^3 \times 3b$

c) $6p^4 \times 4p^6$

d) $r^6 \div r^2$

e) $12c^4 \div 3c^3$

f) $18a^2b \div 9a$

g) $15a^3b^2 \div 3ab$

h) $(2a^2b)^3$

4 Simplify…

a) $\dfrac{8ab^2}{4b}$

b) $\dfrac{16a^3b^2}{ab^2}$

c) $\dfrac{(4ab)^2}{2ab^2}$

d) $\dfrac{12bc^2 \times 3ab^2}{4abc}$

5 Simplify…

a) $12x + 6x$ **b)** $12x - 6x$ **c)** $12x \times 6x$ **d)** $12x \div 6x$ **e)** $4p^2 - 11p^2$

f) $19ab - 13bc + 2ab + 16bc$ **g)** $13ab + 4a^2b - 2ab^2 + 3a^2b - 4ab + 10ab^2$

6 Simplify…

a) $a^4 \times a^3 \times a^2$ **b)** $2x^2 \times 3x \times 5x^5$ **c)** $16x^4 \div 16x^2$ **d)** $20a^3b^2c \div 10a^2b$ **e)** $(4a^2)^3$ **f)** $(3a^2b)^4$

30   ALGEBRA   Revision Guide Reference: Page 34 & 35   *Lonsdale* REVISION GUIDES

Substitution

1 **If p = 2, q = 5 and r = -4, find the value of …**

a) $2p + 3q$ ....

b) $2(p + q)$ ....

c) $2pq$ ....

d) p^2q ....

e) $pq - q^2$ ....

f) pqr ....

g) $p^3 + r^2$ ....

h) $\dfrac{4p}{r}$ ....

i) $p^2q^2 + \dfrac{r}{p}$ ....

j) $\dfrac{p}{q} + \dfrac{r}{q}$ ....

2 **If $x = \frac{1}{2}$, y = $\frac{1}{3}$ and z = -2, find the value of …**

a) $x + y$ ....

b) xz ....

c) xz^2 ....

d) $\dfrac{1}{x} + z$ ....

3 **Find the value of …**

a) $3x - 7$ when $x = -3$ ....

b) $4(x^2 - 1)$ when $x = -3$ ....

c) $4(x - 1)^2$ when $x = 5$ ....

d) $(x + 2)(x - 3)$ when $x = 6$ ....

e) $(x^2 - 5)(x + 8)$ when $x = -5$ ....

---

4 **If e = 3, f = 8, g = -4 and h = $\frac{1}{4}$, find the value of …**
  a) ef **b)** fg **c)** gh **d)** $e + f + g$ **e)** $f + g + h$ **f)** $e^2 - f$ **g)** g^2h **h)** $f \div g$ **i)** $e^3 + g$ **j)** $2f \div g^2$ **k)** $\frac{1}{e} + h$ **l)** $e^2 + f + g^2$

5 **Find the value of …**
  a) $5x^2 - 3$ when $x = -4$ **b)** $5(x - 3)$ when $x = -4$ **c)** $5x^2 - 3$ when $x = -2$ **d)** $(x + 3)(2x - 1)$ when $x = 2$
  e) $(x^2 + 3)(x - 5)$ when $x = -1$

6 **If m = 3 x 10^4 and n = 5 x 10^3, find the value of…**

  a) mn **b)** $m + n$ **c)** $\dfrac{mn}{m+n}$

  Give your answers in standard form to 2 significant figures.

Brackets and Factorisation

1 **Expand and simplify ...**

a) $4(2x + 1)$

b) $3(3r - 7)$

c) $2m(3m + 2)$

d) $2p(4 - p)$

e) $6r(r^2 - 3)$

f) $10x(4x^2 - 3x)$

g) $5(2y + 4) + 3$

h) $5x(2 - 3x) + 7x$

i) $11x(5 - 3x^2) - 9x$

j) $5(3x + 2) + 4(x - 3)$

k) $5(6x + 1) + 3(4x - 2)$

l) $5x(2x - 3) - 4(3x - 1)$

m) $(x + 2)(x + 5)$

n) $(2x + 3)(x + 4)$

o) $(3x - 2)(2x + 1)$

p) $(4x - 1)(x - 5)$

q) $(x + 3)(x - 3)$

r) $(4x - 1)^2$

2 **Factorise the following expressions:**

a) $5x + 10$

b) $4x - 8$

c) $6x + 10y$

d) $6x + 3x^2$

e) $10x^2 - 5x$

f) $6x^2y - 10xy$

g) $4p^2q^3r + 6pqr$

h) $x(3y + 2) + z(3y + 2)$

i) $x(4y + 3) + (4y + 3)^2$

---

3 **Expand and simplify ...**
  a) $6(4x - y)$ **b)** $3x(2x + 5)$ **c)** $4x^2(2y - x)$ **d)** $6x(3 - x^2)$ **e)** $4(2x + 3) + 5(x - 7)$ **f)** $6x(2x - 3) + 4(x + 5)$
  g) $10x^2(4x + 3) - 6x(2x - 1)$ **h)** $(4x + 2)(3x + 2)$ **i)** $(5x + 6)(5x - 6)$ **j)** $(4x^2 + 3x)(2x + 2)$ **k)** $(2x^3 - 3)(4 - 3x^2)$

4 **Factorise the following expressions:**
  a) $9x - 15$ **b)** $9x + 9$ **c)** $20x^2 - x$ **d)** $4x - 20x^3$ **e)** $20x^2y^2 + 36xy$ **f)** $4(3x - 5) + y(3x - 5)$

5 **a)** Expand and simplify $(a + b)^2$ **b)** Without using a calculator work out the value of $8.8^2 + 2 \times 8.8 \times 1.2 + 1.2^2$

Solving Linear Equations 1 & 2

1 Solve the following equations:

a) $5x = 35$

b) $3x + 4 = 16$

c) $5x + 8 = 23$

d) $2(x + 2) = 12$

e) $5(x - 3) = 10$

f) $4(5 + 3x) = 14$

g) $\dfrac{x}{3} = 6$

h) $\dfrac{x}{2} + \dfrac{x}{6} = 10$

i) $\dfrac{x + 2}{5} = 3$

j) $5x = 2x + 9$

k) $7x = 15 - 3x$

l) $2x = x - 8$

m) $11x + 3 = 3x + 7$

n) $5 + 7x = 23 + 3x$

o) $10x - 6 = 3x + 15$

p) $8(3x - 2) = 20$

q) $5(x + 7) = 3(9 + x)$

r) $\dfrac{x + 1}{2} + \dfrac{x - 3}{4} = 2$

2 **a)** Joan thinks of a number.

If you multiply it by 3 and add 12 the answer is 27.

Let the number Jean thinks of be x.

Form an equation and solve it.

b) Jim thinks of a number.

If you divide it by 4 and then add 11 the answer is 13.

Let the number Jim thinks of be x.

Form an equation and solve it.

Solving Linear Equations 1 & 2

3 Maggie is x years old.   Nigel is twice Maggie's age.   Helen is 4 years older than Nigel.

a) Write down an expression in terms of x for their combined age.

....

....

....

b) Their total combined age is 64 years.

Form an equation and solve it to find Helen's age.

....

....

....

4 ABC is a triangle:

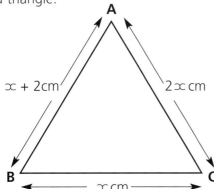

$x + 2$ cm $2x$ cm

x cm

a) Write down an expression in terms of x for the perimeter of the triangle.

....

....

b) The perimeter of triangle ABC is 50cm.

Form an equation and solve it to find x.

....

....

....

5 ABCD is a rectangle:

A   B

6cm

D ← x cm → ← 5cm → C

a) Write down an expression in terms of x for the area of the rectangle.

....

....

b) The area of rectangle ABCD is 72cm^2.

Form an equation and solve it to find x.

....

....

....

---

6 Solve the following equations to find the value of x:

a) $2x + 5 = 17$ **b)** $15 + 3x = 3$ **c)** $4x = x + 18$ **d)** $10x = 7x - 15$ **e)** $4x - 3 = 6x + 12$ **f)** $9x + 7 = 4x - 13$

g) $14 + 5x = 7x + 2$ **h)** $23 - 8x = x - 4$ **i)** $40x - 36 = 7x + 30$ **j)** $4(x + 5) = 36$ **k)** $7(2x - 3) = 14$ **l)** $6 = 4(3x - 9)$

m) $16 = 4(11 - 5x)$ **n)** $5(x + 1) = 2(x + 7)$ **o)** $11(3x + 2) = (6x - 5)$ **p)** $9(2x + 4) - 4(5x + 8) = 0$

7 Janice thinks of a number. If you subtract 6 from it and then multiply by 3 the answer is 21.
Let the number that Janice thinks of be x. Form an equation and solve it to find the value of x.

8 For each of the following form an equation and solve it to find x.

a) 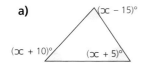 $(x - 15)°$

$(x + 10)°$ $(x + 5)°$

b) $(3x + 20)°$ 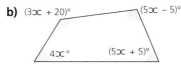 $(5x - 5)°$

$4x°$ $(5x + 5)°$

c) 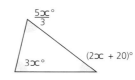 $\frac{5x}{3}°$

$(2x + 20)°$

$3x°$

Formulae 1 & 2

1 Rearrange the following formulae to make x the subject:

a) $x + 4y = 3$

b) $6x + 7y = 50$

c) $7x - 5 = 3y$

d) $6y = 10 - 3x$

e) $\dfrac{x + 3y}{4} = 5$

f) $\dfrac{4x - 3}{y} = 8$

g) $5x + 3y = 3(6 - x)$

h) $4x^2 = 3y$

i) $\dfrac{x^2}{3} = 6y$

j) $5x^2 + 3 = 7y$

2 The volume of a cone is given by the formula: volume $= \frac{1}{3}\pi r^2 h$.

Calculate the volume of a cone in cm^3 if r = 4cm, h = 10cm and π = 3.

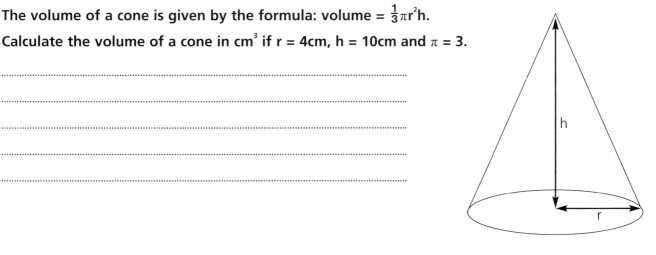

Formulae 1 & 2 (cont)

3 The distance travelled (s) by an object depends on its initial speed (u), its final speed (v) and the time of travel (t). It is given by the formula:

$$s = \left(\frac{u + v}{2}\right) t$$

Calculate the distance travelled in metres if initial speed = 4m/s, final speed = 12m/s and time = 5.5s

....

....

4 The formula that converts a temperature reading from degrees Celsius (°C) into degrees Fahrenheit (°F) is:

$$F = \frac{9}{5} C + 32$$

What is the temperature in degrees Celsius if the temperature in degrees Fahrenheit is 212°F?

....

....

....

5 The circumference of a circle is given by the formula C = 2πr, where r is the radius.

The area of a circle is given by the formula $A = \pi r^2$.

a) Generate a formula for the area of a circle in terms of its circumference, C.

....

....

....

b) Use your formula to work out the area of a circle in cm^2, which has a circumference of 40cm and π = 3.14. Give your answer to two significant figures.

....

6 Rearrange the following formulae to make p the subject:
   a) $4p - 3 = 4q$ **b)** $4 - 6p = q$ **c)** $3(2p + 5) = 4p + 11q$ **d)** $3p^2 + 4 = 8q$ **e)** $\dfrac{3p^2 - 6}{4} = 2q$

7 The area of a parallelogram is given by the formula: Area = length x height. Calculate the area, in m^2, of a parallelogram that has length = 90cm and height = 1.2m.

8 The area, A, of a circle is given by the formula: $A = \pi r^2$, where r is the radius. Calculate the radius if area = $100cm^2$ and π = 3.14. Give your answer to two significant figures.

9 A garden centre buys shrubs at wholesale prices. They calculate the sale price (s), at which they are sold to customers, by increasing the wholesale price (w) by 50% and adding £2.50 per shrub.
   a) Generate a formula for calculating the sale price of a shrub. **b)** If the wholesale price of a shrub is £4.00, use your formula to calculate its sale price.

Factorising Quadratic Expressions

1 **Factorise the following quadratic expressions:**

a) $x^2 + 6x + 8$

b) $x^2 + 7x + 10$

c) $x^2 + 6x + 9$

d) $x^2 - 9x + 20$

e) $x^2 + 9x - 10$

f) $x^2 + 8x - 20$

g) $x^2 - x - 12$

h) $x^2 - 10x - 24$

i) $x^2 - 9$

j) $x^2 - 64$

k) $x^2 - 100$

l) $x^2 - 144$

---

2 **Factorise the following quadratic expressions:**

a) $x^2 + 5x + 6$ **b)** $x^2 - 5x + 6$ **c)** $x^2 + 5x - 6$ **d)** $x^2 - 5x - 6$ **e)** $x^2 + x - 30$ **f)** $x^2 + 31x + 30$ **g)** $x^2 - 81$ **h)** $x^2 - 169$

Solving Quadratic Equations

1 Solve the following quadratic equations:

a) $x^2 + 11x + 24 = 0$

b) $x^2 + 7x - 30 = 0$

c) $x^2 - 16x - 36 = 0$

d) $x^2 - 8x + 12 = 0$

e) $x^2 - 9 = 0$

f) $x^2 - 25x = 0$

2 This open box has a square base.

a) Show that the outside surface area of the open box, A, is given by $A = x^2 + 8x$.

b) If the outside surface area is 65cm^2, find the value of x.

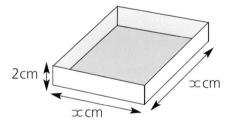

2cm

x cm

x cm

3 Solve the following equations to find the value of x:

a) $x^2 + 12x + 11 = 0$ **b)** $x^2 + 4x - 21 = 0$ **c)** $x^2 - 12x + 27 = 0$ **d)** $x^2 - 14x + 40 = 0$ **e)** $x^2 + 3x - 40 = 0$
f) $x^2 + 9x - 22 = 0$ **g)** $x^2 - 9x - 22 = 0$ **h)** $x^2 - x - 42 = 0$ **i)** $x^2 + 9x = 0$ **j)** $x^2 - 6x = 16$ **k)** $x^2 + 15 = 8x$
l) $3x = 70 - x^2$ **m)** $x^2 - 1 = 0$ **n)** $x^2 = 100$

4 A rectangular box has the following dimensions:

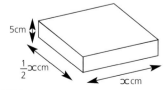

5cm

$\frac{1}{2}x$ cm

x cm

a) Show that the surface area of the box, A, is given by $A = x^2 + 15x$.
b) If the surface area, A, is 34cm^2 find the value of x.
c) Show that the volume of the box, V, is given by $V = 2.5x^2$
d) Calculate the volume of the box using your answer from part b).

Trial and Improvement

1 The equation $x^3 - x = 15$ has a solution which lies between 2 and 3. Using trial and improvement complete the table to solve the equation and find x. Give your answer correct to 1 decimal place.

x	$x^3 - x$	Comment
2	$2^3 - 2 = 8 - 2 = 6$	Less than 15
3	$3^3 - 3 = 27 - 3 = 24$	More than 15

Answer: ....

3 A rectangular box has the following dimensions:

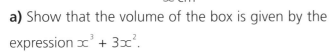

x cm

$x + 3$cm

x cm

a) Show that the volume of the box is given by the expression $x^3 + 3x^2$.

....

....

b) The volume of the box is 40cm^3. Using trial and improvement, find x which has a value that lies between 2 and 3. Give your answer to 1 decimal place.

2 The equation $x^3 + 2x = 40$ has a solution which lies between 3 and 4. By trial and improvement calculate a solution to 2 decimal places.

x	$x^3 + 2x$	Comment

Answer: ....

x	$x^3 + 3x^2$	Comment

Answer: ....

4 **a)** The equation $x^3 + 10x = 24$ has one solution which lies between 1 and 2. Using trial and improvement, find the solution to 1 decimal place.
b) The equation $x^3 - 6x = 65$ has one solution which lies between 4 and 5. Using trial and improvement, find the solution to 2 decimal places.

5 A rectangular box has the following dimensions: length = $2x$ cm, width = x cm and height = $x + 3$cm.
a) Show that the volume of the box is given by the expressions $2x^3 + 6x^2$
b) The volume of the box is 50cm^3. Using trial and improvement find x, which has a value that lies between 2 and 3. Give your answer to 1 decimal place.

1 **a)** The first four numbers of a sequence are:

1, 3, 7, 15...

The rule to continue this sequence of numbers is:

b) The first four numbers of a sequence are:

3, 4, 6, 10...

The rule to continue this sequence of numbers is:

Multiply the previous number by 2 and then add 1

Subtract 1 from the previous number and then multiply by 2

i) What are the next two numbers in the sequence?

....

ii) The following sequence obeys the same rule:

-2, -3, -5, -9...

What are the next two numbers in this sequence?

....

i) What are the next two numbers in the sequence?

....

ii) The following sequence obeys the same rule:

1, 0, -2, -6...

What are the next two numbers in this sequence?

....

2 **a)** The first four square numbers are 1, 4, 9 and 16. What are the next four numbers in the sequence?

....

b) The first four triangular numbers are 1, 3, 6 and 10. What are the next four numbers in the sequence?

....

3 **The nth term of a sequence is 5n + 8.**

a) What is the value of the 3rd term?

....

b) What is the value of the 10th term?

....

4 **The nth term of a sequence is 2n – 11.**

a) What is the value of the 4th term?

....

b) What is the value of the 16th term?

....

5 **The nth term of a sequence is 2n – 9.**

a) Which term has a value of 19?

....

b) Which term has a value of -5?

....

6 **The nth term of a sequence is 7n + 6.**

a) Which term has a value of 69?

....

b) Which term has a value of 90?

....

7 The first four terms of a sequence are:

3, 5, 7, 9, ...

Write down a formula for the nth term of this sequence and add a further 5 terms to the sequence.

....

....

....

8 The first four terms of a sequence are:

6, 4, 2, 0, ...

Write down a formula for the nth term of this sequence and add a further 5 terms to the sequence.

....

....

....

9 Here is a sequence of diagrams made up of squares:

Diagram 1   Diagram 2   Diagram 3   Diagram 4

a) Write down a formula for the number of squares (s) in terms of diagram number (n).

....

....

....

b) How many squares would there be in Diagram 8?

....

c) Which number diagram would have 49 squares?

....

....

10 Here is a sequence of diagrams made up of circles.

Diagram 1   Diagram 2   Diagram 3   Diagram 4

a) Write down a formula for the number of circles (c) in terms of diagram number (n).

....

....

b) How many circles would there be in Diagram 15?

....

c) Which number diagram would have 81 circles?

....

....

---

11 Draw the next two diagrams of the following sequences:

a)

b)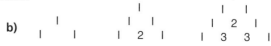

12 The nth term of a sequence is $\frac{(4n - 7)}{3}$ **a)** What is the value of the 10th term? **b)** Which term has a value of 27?

13 The first four terms of a sequence are: 15, 11, 7, 3,...
   a) Write down a formula for the nth term of this sequence. **b)** What is the value of **i)** the 10th term? **ii)** the 100th term?

14 Write down a formula for the nth term of the sequence 2, 5, 10, 17...

Plotting Points

1 Write down the coordinates of all the plotted points below.

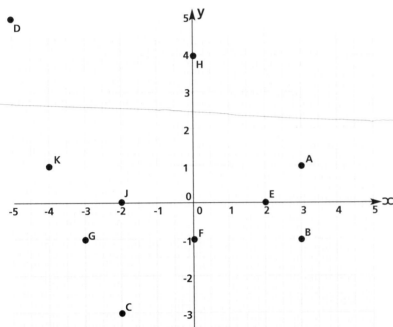

A (..................... , ...)

B (..................... , ...)

C (..................... , ...)

D (..................... , ...)

E (..................... , ...)

F (..................... , ...)

G (..................... , ...)

H (..................... , ...)

I (..................... , ...)

J (..................... , ...)

K (..................... , ...)

L (..................... , ...)

2 Use the axes below to help you answer the following questions:

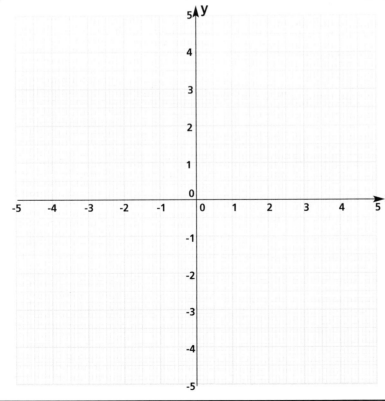

a) The coordinates of 3 points are A (3,0), B (0,-3), C (-3,0). What are the coordinates of point D if ABCD is a square?

....

b) The coordinates of 3 points are P (4,2), Q (3,-2), R (-2,-2). What are the coordinates of point S if PQRS is a parallelogram?

....

c) The coordinates of 3 points are E (1,5), F (4,4), G (1,-3). What are the coordinates of point H if EFGH is a kite?

....

3 The coordinates of 3 points are A (2,3), B (2,-2), C (-3,-2). What are the coordinates of point D if ABCD is a square?

4 The coordinates of 3 points are P (-4,0), Q (3,4), R (4,0). What are the coordinates of point S if PQRS is a kite?

Graphs of Linear Functions 1

1 **On the axes provided, draw and label the graphs of the following linear functions for values of x between -2 and 2:**

a) $y = 2x$

x	-2	0	2
y	-4	0	4

b) $y = 2x - 1$

x	-2	0	2
y	-5		

c) $y = x + 2$

x			
y			

d) $y = -2x$

x			
y			

e) $y = -x - 3$

x			
y			

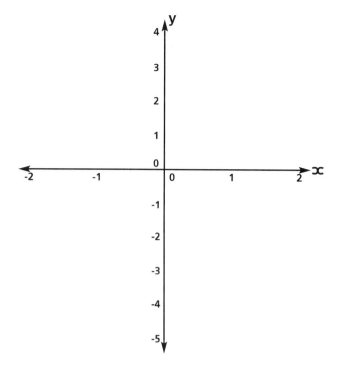

2 **a)** Make y the subject of the following function:

$2y - x = 8$

....

....

b) On the axes provided, draw and label the graph of the rearranged function from part a).

x			
y			

c) A point, which lies on the line of the graph that you have just drawn, has the coordinates (5,p). Calculate the value of p.

....

....

d) Another point that lies on the line of the drawn graph has the coordinates (q,0). Calculate the value of q.

....

....

---

3 **a)** Make y the subject of the following function: $y - 3x = -4$
   b) Draw the graph of the rearranged function for values of x between -3 and 3.
   c) Use the graph to calculate the value of y if x = 1.5   **d)** Use the graph to calculate the value of x if y = 3.2

4 **a)** Draw graphs of $y = 3x$ and $y = x + 5$ for values of x between 0 and 3 on the same set of axes.
   b) What is the x coordinate of the point where the two lines cross?

5 **Draw the graphs of $y = 2x - 1$ and $y = -2$. Find the coordinate of the point where they cross.**

Graphs of Linear Functions 2

1 Find the gradient and intercept of the following linear functions:

a) $y = 2x + 1$

....

....

....

b) $y = 3x - 1$

....

....

....

c) $x + y = 3$

....

....

....

d) $4y = 7 - 8x$

....

....

....

e) $2y - 2x = 9$

....

....

....

f) $2x = y - 2$

....

....

....

2 On the axes below are the graphs of 4 lines whose equations can be found in question 1. What is the equation of each line?

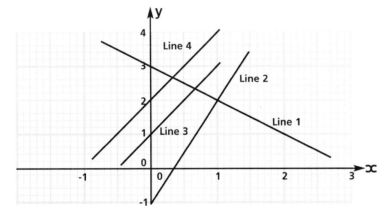

Line 1: ....

Line 2: ....

Line 3: ....

Line 4: ....

3 What is the equation of the line which crosses the y-axis at ...

a) (0,1) and is parallel to $y = 2x$? ....

b) (0,1) and is parallel to $y = -2x$? ....

c) (0,-2) and is parallel to $y = 2x$? ....

d) (0,-2) and is parallel to $y = x + 2$? ....

e) (0,3) and is parallel to $y = -x - 3$? ....

f) (0,-3.5) and is parallel to $y = -x$? ....

---

4 Find the gradient and intercept of the following linear functions:
a) $y = -3x - 3$ **b)** $4y = 3x + 8$ **c)** $2y - x = 3$ **d)** $x - 2y = 3$ **e)** $2y - 3 = x$ **f)** $\dfrac{y - 2x}{3} = 5$ **g)** $\dfrac{y + 4x}{5} = 1$

5 What is the equation of the line which crosses the y-axis at **a)** (0, 4) and is parallel to $y = x - 4$, **b)** (0, 0) and is parallel to $y = x - 4$

6 Which of the following linear functions would produce parallel lines if drawn on the same axes:
i) $y = 2x + 3$ **ii)** $y + 2x = 3$ **iii)** $2y - 4x = 7$ **iv)** $y - 6 = 2x$

Graphs of Linear Functions 3

1 Find the gradient, intercept and equation of the following lines.

a)

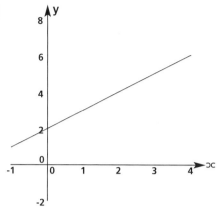

....
....
....

b)

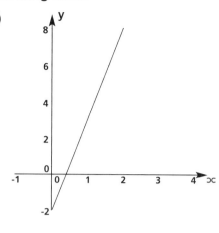

....
....
....

c)

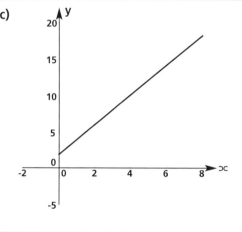

....
....
....

d)

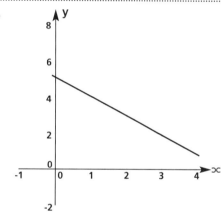

....
....
....

2 A sketch of the line 2y - 6 = **x** is shown.

The line crosses the axes at A and B.

a) What is the gradient of the line AB?

....

b) Calculate the coordinates of A and B.

....
....

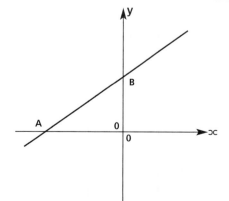

3 Plot the following points: A(0,1), B(4,5), C(1,4) and D(4,1).
   a) Calculate the gradient of **i)** line AB, **ii)** line CD **b)** What is the equation of **i)** line AB, **ii)** line CD?

4 A line has intercept c = +2. A point with coordinates (4,4) lies on the line. What is the equation of the line?

5 A line has intercept c = +5. A point with coordinates (4,1) lies on the line. What is the equation of the line?

Three Special Graphs

1 On the axes below are the graphs of 6 lines. What is the equation of each line?

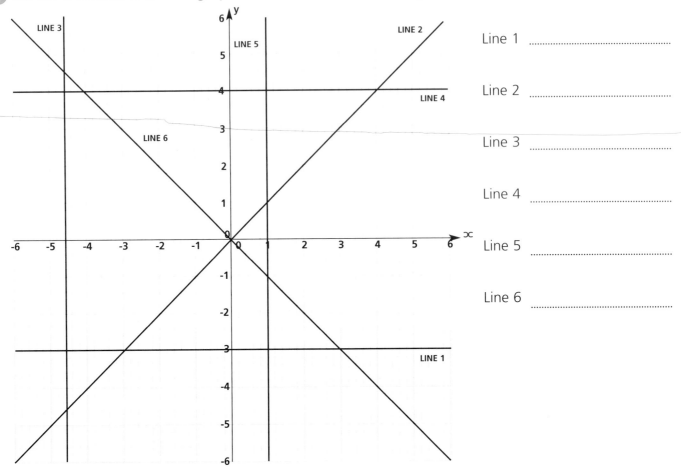

Line 1 ....

Line 2 ....

Line 3 ....

Line 4 ....

Line 5 ....

Line 6 ....

2 **a)** Two points which lie on the line of a drawn graph have coordinates (-3,0) and (4,0).

What is the equation of the line?..

b) Two points which lie on the line of a drawn graph have coordinates (-1,5) and (-1,3).

What is the equation of the line?..

c) Two points which lie on the line of a drawn graph have coordinates (-4,4) and (4,-4).

What is the equation of the line?..

3 **What is the equation of the line that passes through the point with coordinates (3,3) and has a gradient of 1?**..

4 **a)** What are the coordinates of the point of intersection of the two lines $y = x$ and $y = -3$?

....

b) What are the coordinates of the point of intersection of the two lines $y = 6$ and $x = -2$?

....

5 **What is the equation of the line that is parallel to the x-axis and passes through the point with coordinates (2.5, 3.6)?**

6 **What is the equation of the line that is perpendicular to the x-axis and passes through the point with coordinates (-2.8, 1.2)?**

Linear Inequalities 1 & 2

1 Solve the following inequalities and draw a number line for each one:

a) $x + 2 > 11$

b) $8 > x - 9$

c) $2x + 5 \leqslant 15$

d) $13 \geqslant 5 + 4x$

e) $7x - 2 \leqslant 2x + 13$

f) $6 + 3x < 21 - 2x$

2 Solve the following inequalities:

a) $12 + 3x < 6$

b) $14 > 5 - 3x$

c) $5(4x + 7) \geqslant 15$

d) $3(2x + 5) \geqslant 24$

e) $6(4 - x) \leqslant 9$

f) $8 \geqslant 3 - 2x$

3 Draw, label and shade the region represented by each inequality on the grids provided.

a) $x \geqslant 1$

b) $y < -2$

c) $y > -x + 2$

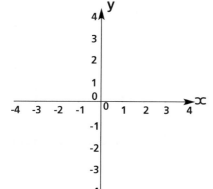

d) $y \leqslant 2x - 1$

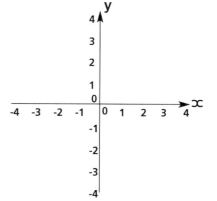

Linear Inequalities 1 & 2 (cont)

4 a) On the grid, draw and shade the region that satisfies these three inequalities:

$x \leqslant 3$, $y \leqslant 4$ and $x + y \geqslant 4$

Label the region A.

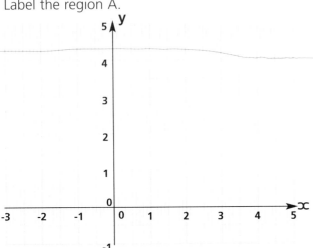

b) B is a point within region A, which has coordinates that are both integers.

What are the coordinates of B?

....

5 a) On the grid, draw and shade the region that satisfies these three inequalities:

$x < 4$, $y \geqslant 1$ and $x < y$

Label the region A.

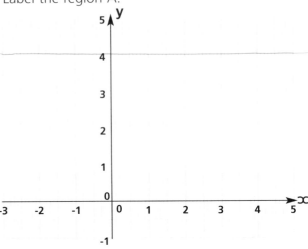

b) B is a point within region A, which has coordinates that are both integers.

What are the coordinates of B?

....

6 On the grid, draw and shade the region that satisfies these four inequalities:

$x > \text{-}2$, $y \geqslant 0$, $y \geqslant 0.5x$, $x + y < 3$

Label the region A.

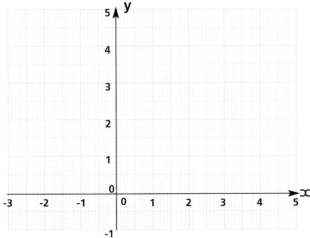

7 On the grid, the region labelled A satisfies four inequalities, what are they?

....

....

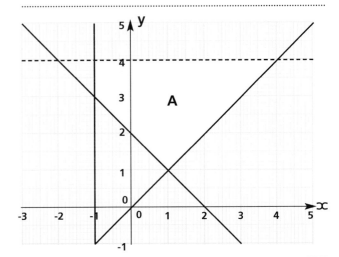

8 Solve the following inequalities and for each one draw a number line. **a)** $4x \geqslant 20$ **b)** $x + 3 > 5$ **c)** $4x - 3 \leqslant 19$
d) $8x + 5 < 2x - 7$ **e)** $5(2x - 3) < 4x + 9$ **f)** $7(3x - 2) \geqslant 20x - 16$ **g)** $6(3x - 2) < 4(4x - 10)$

9 On a suitable grid, draw and shade the region that satisfies these three inequalities:
$x \geqslant 2$, $y \leqslant 5$ and $y \geqslant x$. Label the region A.

10 On a suitable grid, draw and shade the region that satisfies these four inequalities:
$x > \text{-}1$, $y \geqslant x - 2$, $y > \text{-}1$ and $y \leqslant \text{-}x + 5$. Label the region A.

Simultaneous Equations 1

1 Solve the following simultaneous equations:

a) $2x + y = 8$

$x + y = 5$

b) $2x + 5y = 24$

$3x - 5y = 11$

c) $x - 6y = 17$

$3x + 2y = 11$

d) $4x + 3y = 27$

$x + y = 7$

e) $5x - 2y = 14$

$2x + 3y = 17$

f) $3x + 4y = 13$

$2x - 3y = 3$

2 At break time a student buys two doughnuts and a coffee, which cost her 84p altogether. At lunchtime, the same student buys three doughnuts and two coffees, which cost 138p altogether. Form two equations with the information given and work out the individual price of a doughnut (d) and a coffee (c).

3 Solve the following simultaneous equations:

a) $5x + y = 14$, $3x + y = 10$ **b)** $4x - y = 6$, $x + y = 9$ **c)** $3x + 2y = 8$, $x + y = 2$ **d)** $4x + 6y = 12$, $x + y = 1$

e) $4x + 2y = 13$, $10x - 4y = 1$ **f)** $x - 10y = 2$, $4x + 2y = -13$ **g)** $4x - 3y = -6$, $y - 3x = 7$ **h)** $10x + 2y = 0$, $x - y = 9$

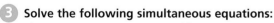

Simultaneous Equations 2

1 a) On the axes below, draw and label the graph of the following equations:

$y = x + 8$ and $y = 2x + 5$

x	0	2	4
y=x+8	8	10	12

x	0	2	4
y=2x+5	5	9	13

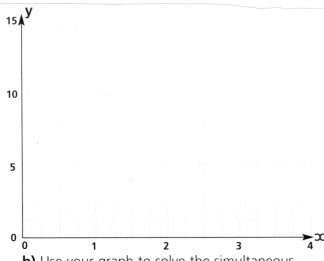

b) Use your graph to solve the simultaneous equations: $y = x + 8$ and $y = 2x + 5$

....

2 a) On the axes below, draw and label the graph of the following equations:

$y = 3x + 10$ and $y = -x + 22$

x	0	2	4
y=3x+10			

x	0	2	4
y=-x+22			

b) Use your graph to solve the simultaneous equations: $y = 3x + 10$ and $y = -x + 22$

....

3 a) On the axes provided draw the graphs of the following lines:

i) $y = -x + 4$

ii) $y = x + 2$

iii) $y = -0.25x + 1$

b) Use your graph to solve the following simultaneous equations:

i) $y = x + 2$ and $y = -0.25x + 1$ ....

ii) $y = x + 2$ and $y = -x + 4$ ....

iii) $y = -x + 4$ and $y = -0.25x + 1$ ....

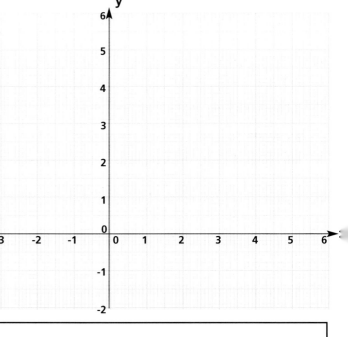

4 a) Draw the graph of $2x + y = 13$ and $3x = y - 3$ for values of x between 0 and 4.
b) Use your graph to solve the simultaneous equations $2x + y = 13$ and $3x = y - 3$

5 a) Draw the graph of $x + y = 6$, $3x - y = 2$ and $x - y = 4$ for values of x between -2 and 6.
b) Use your graph to solve the following simultaneous equations: **i)** $x + y = 6$, $3x - y = 2$ **ii)** $x + y = 6$, $x - y = 4$
iii) $3x - y = 2$, $x - y = 4$

Graphs of Quadratic Funcs. 1 & 2

1 **Below is a table of values for $y = x^2 - 2$.**

x	-2	-1	0	1	2
y	2	-1	-2	-1	2

a) Draw the graph of $y = x^2 - 2$.

b) From your graph find the value(s) of...

i) y when $x = 1.5$

....

ii) x when y = 1.5

....

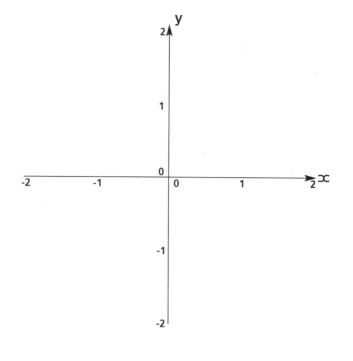

2 **Below is a half completed table of values for**

$y = x^2 + x - 3$.

a) Complete the table.

x	-3	-2	-1	0	1	2
x^2		4		0		4
$+x$		-2		0		2
-3		-3		-3		-3
$y = x^2 + x - 3$		-1		-3		3

b) Draw the graph of $y = x^2 + x - 3$.

c) From your graph find the value(s) of...

i) y when $x = -2.5$

....

ii) x when y = 1.6

....

d) From your graph find the solution to the equation $x^2 + x - 3 = 0$.

....

e) From your graph find the solution to the equation $x^2 + x - 3 = 1$.

....

f) From your graph find the solution to the equation $x^2 + x - 3 = -2$.

....

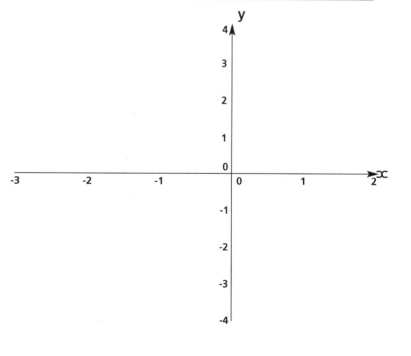

3 Below is a half completed table of values for $y = x^2 + 2x - 4$.

x	-4	-3	-2	-1	0	1	2
x^2		9		1		1	
$+2x$		-6		-2		2	
-4		-4		-4		-4	
$y = x^2 + 2x - 4$		-1		-5		-1	

a) Complete the table.

b) Draw the graph of $y = x^2 + 2x - 4$.

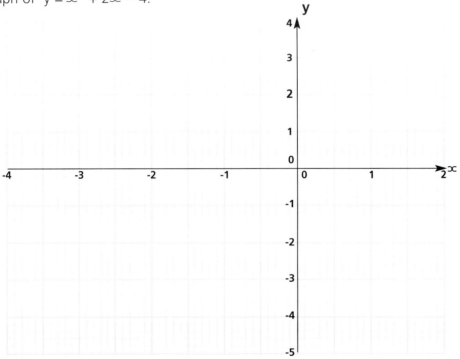

c) From your graph find the solutions to the equations...

i) $x^2 + 2x - 4 = 0$

ii) $x^2 + 2x - 6 = 0$

iii) $x^2 + 2x = 1$

4 **a)** Draw a table of results for $y = x^2 + 4x - 6$ for values of x between -6 and 2.
   b) Draw the graph of $y = x^2 + 4x - 6$.
   c) From your graph find the solutions to the equations: **i)** $x^2 + 4x - 6 = 0$ **ii)** $x^2 + 4x - 9 = 0$ **iii)** $x^2 + 4x + 2 = 0$

5 **a)** Draw the graph of $y = x^2 - 12$ for values of x between -4 and 4.
   b) From your graph find the solutions to the equations: **i)** $x^2 - 12 = 0$ **ii)** $x^2 = 15$ **iii)** $x^2 - 6 = 0$

6 **a)** Draw the graph of $y = x^2 - x - 8$ for values of x between -3 and 4.
   b) From your graph find the solutions to the equations: **i)** $x^2 - x - 8 = 0$ **ii)** $x^2 - x - 8 = -2$ **iii)** $x^2 - x - 8 = 3$

Graphs of Other Functions

1 **Below is a half completed table of values for y = x^3 + 10.**

a) Complete the table.

x	-3	-2	-1	0	1	2
x^3		-8		0		8
+10		+10		+10		+10
y = x^3+10		2		10		18

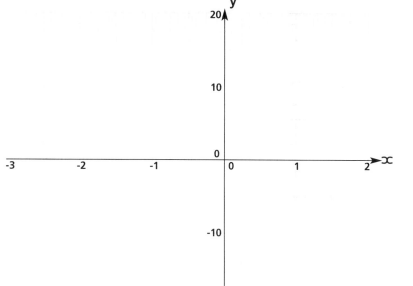

b) On the axes opposite draw the graph of y = x^3 + 10.

c) Use your graph to find ...

i) the value of x, when y = -10

....

ii) the value of y, when x = 1.5

....

2 **Below is a table of values for**

y = $\frac{1}{x}$ with $x \neq 0$.

x	-5	-4	-3	-2	-1	-0.5	-0.$\dot{3}$	-0.25	-0.2
y=$\frac{1}{x}$	-0.2	-0.25	-0.$\dot{3}$	-0.5	-1	-2	-3	-4	-5

x	0.2	0.25	0.$\dot{3}$	0.5	1	2	3	4	5
y=$\frac{1}{x}$	5	4	3	2	1	0.5	0.$\dot{3}$	0.25	0.2

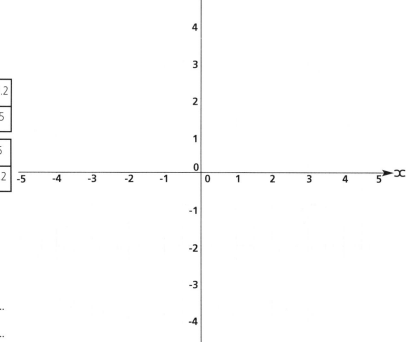

a) On the axes opposite draw the graph of y = $\frac{1}{x}$ with $x \neq 0$.

b) Use your graph to find...

i) the value of x when y = 2.5 ....

ii) the value of y when x = -1.5.............................

---

3 **Below is a half completed table of values for y = x^3 - 2.**

x	-3	-2	-1	0	1	2	3
x^3		-8		0		8	
-2		-2		-2		-2	
y= x^3 − 2		-10		-2		6	

a) Complete the table.
b) Draw the graph of y = x^3 - 2.
c) Use your graph to find the value of x, when y = 0
d) Use your graph to find...
i) the value of x when y = 20
ii) the value of y when x = 1.5

Other Graphs 1

1 Two towns A and B are 100 miles apart.

Mr Brown lives in A and drives to B,

Mr Smith lives in B and drives to A, on the

same day, along the same route.

Using the graph opposite:

a) What time did Mr Smith set off?

....

b) Which motorist completed the journey in

the shortest time? ....

c) Who reached the highest speed and what was it? ....

d) What happened at Y? ....

e) Who stopped and for how long? ....

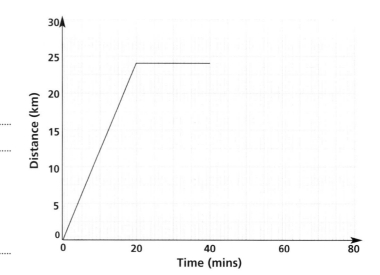

2 **Here is part of a travel graph of Tina's**

journey to the shops and back.

a) Calculate Tina's speed in km/h for the first

20 mins..

....

b) Tina spends 20 mins at the shops, then

travels back home at 48km/h.

Complete her journey on the graph.

....

3 **a)** Plot the following data onto this velocity-time graph.

Time (s)	0	1	2	3	4	5	6
Velocity (m/s)	0	5	10	15	20	20	25

b) Calculate the acceleration of the object

over the first 4 seconds.

....

....

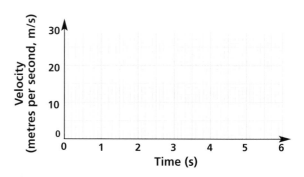

4 An object travels at a constant speed of 20 m/s for the first 6 seconds, then accelerates at 5 m/s² for another 4
seconds. Draw a graph to show the speed of the object over 10 seconds.

5 A train leaves town A for town B at 1pm and maintains a steady speed of 60km/h. At 2pm another train leaves
B for A maintaining a speed of 72km/h. The distance between A and B is 180km. Draw the distance-time graphs
for these trains on the same axes. When do they pass each other?

Other Graphs 2

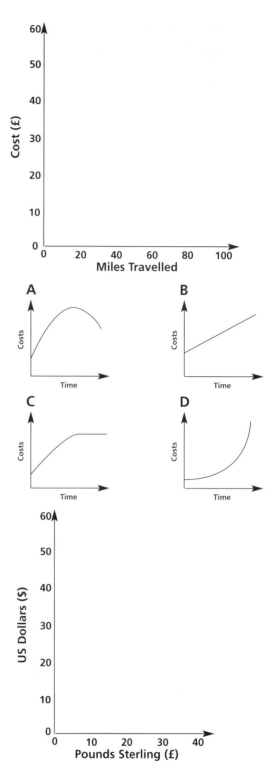

1 **The cost of hiring a car is £30 plus an extra charge of 30p per mile.**

a) Complete this table:

Miles	0	10	20	30	50	100
Cost (£)	30	33				

b) On the axes opposite draw a graph to show the cost of hiring the car for up to 100 miles.

c) Use the graph to estimate how much it would cost to hire a car for a 52 mile journey.

....

2 **These sketch graphs show the cost of running a business over several months. Identify which sketch matches each of these descriptions.**

a) Costs are rising steadily. ....

b) Costs are falling after reaching a peak. ....

c) Costs are rising at an increasing rate. ....

d) Costs have been rising but are now levelling out. ....

3 **a)** If £1 = $1.4 complete the table to convert pounds sterling (£) to US dollars ($).

Pounds (£)	10	20	30	40
US Dollars ($)				

b) On the axes opposite draw the graph to convert pounds (£) to US dollars ($).

c) use your graph to convert $30 into pounds.

....

4 **Three plumbers A, B and C charge as follows: A – Call out charge £20 then £10 per hour extra.
B – Call out charge £30 then £5 per hour extra.   C – Standard charge £50 regardless of time.**

a) Draw a graph for each plumber's charges on the same axes, for up to 5 hrs work.

b) Which plumber is cheapest for a 30 minute job?

c) Which plumber is cheapest for a $2\frac{1}{2}$ hour job?

d) After what time will plumber C become the cheapest?

5 **A tank of water, cuboid in shape, is being drained out at a rate of 10cm depth per minute. If the water is 60cm deep to start with, draw a graph of time against depth.**

Angles 1 & 2

1 For each diagram work out the size of angle p and angle q, giving a reason for your answer.
They are not drawn to scale.

a)

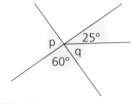

Angle p= ....

Reason: ....

....

Angle q= ....

Reason: ....

....

b)

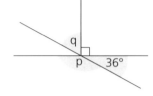

Angle p= ....

Reason: ....

....

Angle q= ....

Reason: ....

....

c)

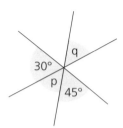

Angle p= ....

Reason: ....

....

Angle q= ....

Reason: ....

....

d)

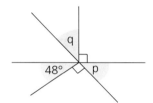

Angle p= ....

Reason: ....

....

Angle q= ....

Reason: ....

....

e)

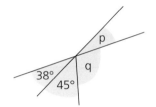

Angle p= ....

Reason: ....

....

Angle q= ....

Reason: ....

....

f)

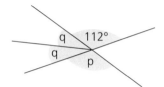

Angle p= ....

Reason: ....

....

Angle q= ....

Reason: ....

....

2 For each diagram work out the size of angle c and angle d giving a reason for your answer.
They are not drawn to scale.

a)

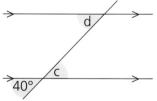

Angle c= ....

Reason: ....

....

Angle d= ....

Reason: ....

....

b)

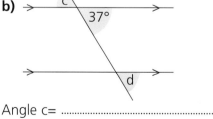

Angle c= ....

Reason: ....

....

Angle d= ....

Reason: ....

....

c)

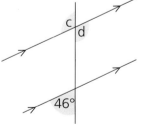

Angle c= ....

Reason: ....

....

Angle d= ....

Reason: ....

....

3 For each diagram work out the size of angle m and angle n giving a reason for your answer. They are not drawn to scale.

a)

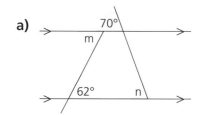

Angle m=...

Reason: ....

....

Angle n= ....

Reason: ....

....

b)

Angle m=...

Reason: ....

....

Angle n= ....

Reason: ....

....

c)

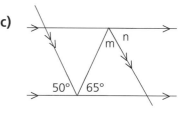

Angle m=...

Reason: ....

....

Angle n= ....

Reason: ....

....

4 Work out the size of the angles marked a, b and c, giving reasons for your answer. The diagram is not drawn to scale.

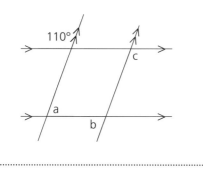

....

....

....

....

5 Work out the size of the angles marked p, q, r and s, giving reasons for your answer. The diagram is not drawn to scale.

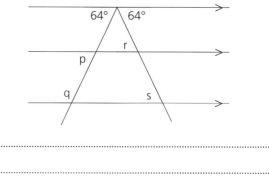

....

....

....

....

6 For each diagram work out the size of x. They are not drawn to scale.

a)

b)

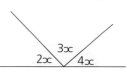

c)

d)

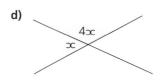

7 For each diagram work out the size of the angles marked giving reasons for your answer. They are not drawn to scale.

a)

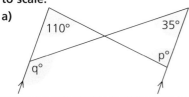

b)

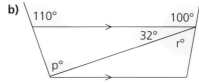

c)

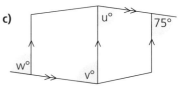

Triangles

1 Draw a diagram and write a short explanation to prove that the interior angles of a triangle add up to 180°.

....
....
....
....
....

2 Draw a diagram and write a short explanation to prove that the exterior angle of a triangle is equal to the sum of the interior angles at the other two vertices.

....
....
....
....
....

3 This diagram has six angles marked a, b, c, d, e and f.

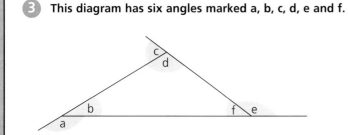

a) Work out the size of angle a in terms of angle b.
b) Work out the size of angle a in terms of angles d and f.
c) What do angles a, c and e add up to?
d) Work out the size of angle a in terms of angles c and e.

Types of Triangle

1 Below are four triangles. For each triangle work out the missing angle x and name the type of triangle giving a reason for your answer. They are not drawn to scale.

a)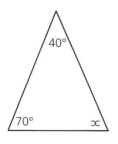

b)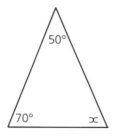

c)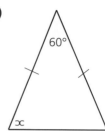

d)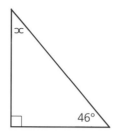

Angle x =........................

Angle x =........................

Angle x =........................

Angle x =........................

Type of Triangle:

Type of Triangle:

Type of Triangle:

Type of Triangle:

....

....

....

....

Reason:

Reason:

Reason:

Reason:

2 For each diagram below work out the size of angle p. They are not drawn to scale.

a)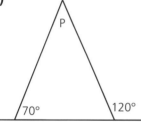

b)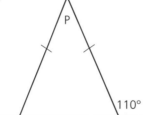

c)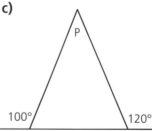

d)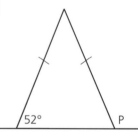

Angle p =........................

Angle p =........................

Angle p =........................

Angle p =........................

3 The diagram (not drawn to scale) shows a right-angled triangle ABC and an isosceles triangle CDE. Work out the size of the angles marked a and b giving reasons for your answers.

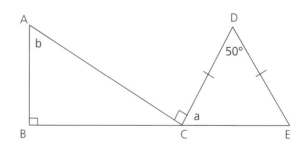

4 The diagram (not drawn to scale) shows an equilateral triangle ABC and a right-angled triangle CDE. Work out the size of the angles marked m and n giving reasons for your answers.

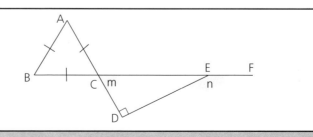

Quadrilaterals

1 Below are four quadrilaterals. For each quadrilateral work out the missing angle x and name the type of quadrilateral, giving a reason for your answer. They are not drawn to scale.

a)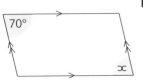

b)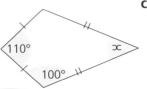

c)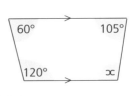

d)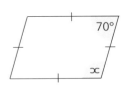

Angle x = ...

Type of Quadrilateral:

....

Reason:

....

....

Angle x = ...

Type of Quadrilateral:

....

Reason:

....

....

Angle x = ...

Type of Quadrilateral:

....

Reason:

....

....

Angle x = ...

Type of Quadrilateral:

....

Reason:

....

....

2 Which type of quadrilateral am I?

a) I have diagonals that are not equal in length but they bisect each other at right angles. They also bisect each of my interior angles. ....

b) I have diagonals that are equal in length and bisect each other. However, they do not bisect each other at right-angles. ....

c) I have diagonals that are equal in length and bisect each other at right angles. They also bisect each of my interior angles. ....

3 Draw a diagram and write a short explanation to prove that the interior angles of a quadrilateral add up to 360°. ....

....

....

....

....

4 Calculate the size of the labelled angles in the following diagrams. They are not drawn to scale.

a)

b)

c)

d)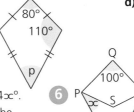

5 **a)** The interior angles of a quadrilateral are $x°$, $2x°$, $3x°$ and $4x°$. Work out the difference in size between the largest angle and the smallest angle in the quadrilateral.

b) The interior angles of a quadrilateral are $x+20°$, $x-30°$, $2x°$ and 110°. Work out the size of x.

6 PQRS is a rhombus and PQRT is a kite. Calculate the value of x.

Irregular & Regular Polygons

1 For each diagram below work out the size of angle p. They are not drawn to scale.

a)

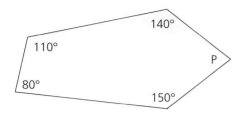

b)

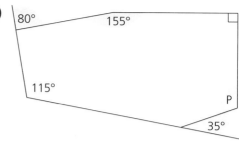

....

....

....

2 **a)** What do the exterior angles of an irregular pentagon add up to? ....

b) What do the exterior angles of a regular pentagon add up to? ....

3 An irregular polygon has exterior angles equal to 110°, 94°, 88° and 68°.

a) How many sides does the polygon have? ....

b) What is the sum of the interior angles of this irregular polygon?

....

....

4 An irregular polygon has interior angles equal to 110°, 155°, 135°, 95°, 140° and 85°.

a) How many sides does the polygon have? ....

b) What is the sum of the exterior angles of this irregular polygon?

....

5 An irregular pentagon has four exterior angles equal to 47°, 113°, 55° and 71°. What is the size of the fifth exterior angle?

....

6 **a)** How many sides does a polygon have if the sum of its interior angles is twice the sum of its exterior angles? ....

....

b) How many sides does a regular polygon have if each interior angle is equal to each exterior angle?

....

c) How many sides does a polygon have if the sum of its interior angles is equal to the sum of its exterior angles? ....

Irregular & Regular Polygons

7 **A regular polygon has five sides.**

Calculate the size of a) each exterior angle and

b) each interior angle.

....

....

....

....

....

....

8 **A regular polygon has six sides.**

Calculate the size of a) each exterior angle and

b) each interior angle.

....

....

....

....

....

....

9 **The diagrams below show regular polygons, centre 0. For each polygon work out the size of the angles marked a and b.**

a)

b)

....

....

....

....

....

....

....

....

10 **Jon reckons that it is possible to have a regular polygon where each interior angle is equal to 150° and each exterior angle is equal to 40°. Is he correct? Explain.**

....

....

11 **A polygon has interior angles equal to $x°$, $x+20°$, $2x°$, $2x-10°$ and 50°.**
a) How many sides does this polygon have?
b) Work out the size of x.
c) What is the size of each exterior angle?

12 **A regular polygon has each interior angle equal to $1\frac{1}{2}$ times each exterior angle.**
a) What is the size of each interior angle?
b) What is the size of each exterior angle?
c) How many sides does this regular polygon have?

13 **A regular polygon has each interior angle equal to 170°. How many sides does it have?**

Congruence & Similarity

1 Triangle A and triangles, P, Q, R, S and T are not drawn accurately.

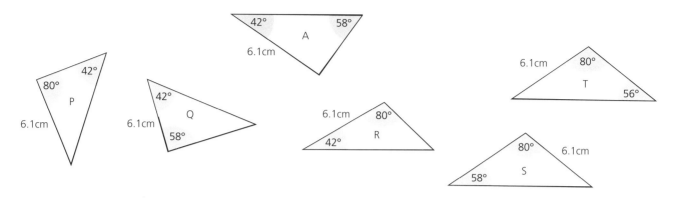

Which two of triangles P, Q, R, S and T are congruent to triangle A?

....

2 a) These two triangles are congruent.

They are not drawn to scale.

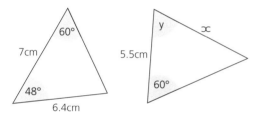

What is x?...

What is y? ....

b) These two quadrilaterals are congruent.

They are not drawn to scale.

What is x?...

What is y? ....

3 On the isometric grid below draw the tessellation formed using regular hexagons of side 1cm.

4 Here are three triangles (not drawn to scale). Which two triangles are congruent? Explain.

5 A regular hexagon will form a tessellation pattern but a regular pentagon does not. Explain why.

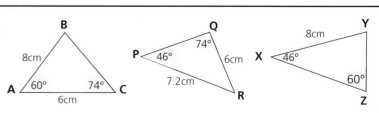

Similar Figures

1 Here are four triangles (not drawn to scale).

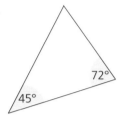 A

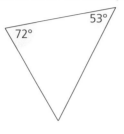 B

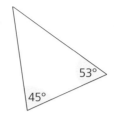 C

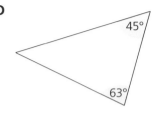 D

a) Pam thinks that triangles B and C are similar. Is she correct? Explain.

....

....

b) Ian thinks that triangles A and D are similar. Is he correct? Explain.

....

....

2 These two triangles are similar. Calculate the length of...

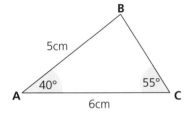

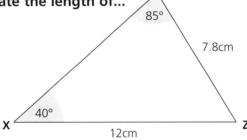

a) XY ....

b) BC ....

3 In the diagram below BC is parallel to DE. AB = 3 cm, AC = 4cm, BC = 5cm and CE = 2cm.

The diagram is not drawn to scale.

a) Calculate the length of DE.

....

....

....

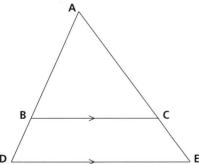

b) Calculate the length of BD.

....

....

....

4 Here are three triangles (not drawn to scale).
Which two triangles are similar? Explain.

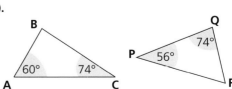

Pythagoras' Theorem 1 & 2

1 Use Pythagoras' Theorem to calculate the unknown side in each of the following triangles.

They are not drawn to scale. If need be, give your answer to 1 decimal place.

a)

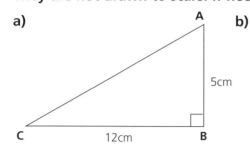

....

....

....

....

b)

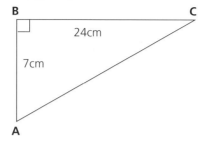

....

....

....

....

c)

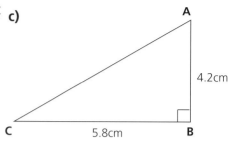

....

....

....

....

d)

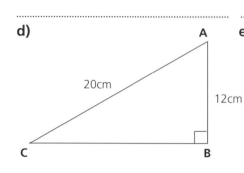

....

....

....

....

e)

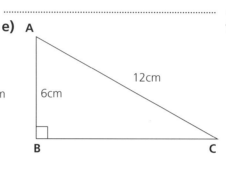

....

....

....

....

f)

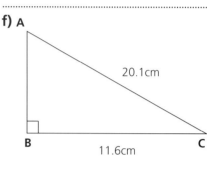

....

....

....

....

2 Which of the following triangles is not a right-angled triangle? Show all your workings.

They are not drawn to scale.

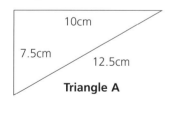

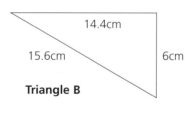

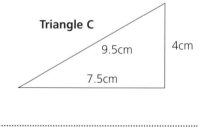

....

....

....

3 Calculate the height of the isosceles triangle alongside, whose sides measure
10cm, 10cm and 5cm. Give your answer to 2 significant figures.
The triangle is not drawn to scale.

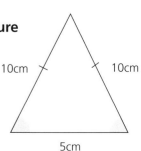

....

....

4 Look at the diagram below (not drawn to scale). Use Pythagoras' Theorem to calculate...

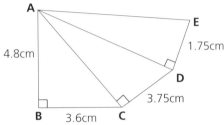

a) the length of AC.

....

....

....

b) the length of AE to 1 decimal place.

....

....

....

....

....

5 The diagram below shows the position of three villages. It is not drawn to scale. Use Pythagoras' Theorem to calculate the direct distance from Lampton to Campton to 1 decimal place.

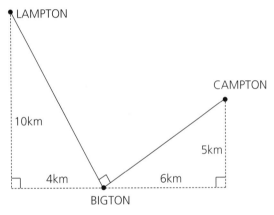

....

....

....

....

....

....

....

....

6 The diagram opposite shows a right-angled triangle ABC.
Use Pythagoras' theorem to calculate...
a) the length of AB to 1 decimal place if BC = 80cm and AC = 1.3m.
b) the length of AC to 1 decimal place if AB = 10cm and the area of triangle ABC is 25cm².
c) the length of AB to 1 decimal place if AB = BC and AC = 14.2cm.

7 Triangle ABC has AB = 1.75cm, BC = 6cm and AC = 6.25cm. Is triangle ABC a right-angled triangle? Explain your answer.

8 A square has an area of 36cm². Calculate the length of its diagonal to 1 decimal place.

9 A rectangle has an area of 36cm² and the length of its base is twice that of its height.
Calculate the length of its diagonal.

10 A 4.5m ladder leans against a wall. The foot of the ladder is 1.5m from the base of the wall.
a) How high up the wall does the ladder reach? Give your answer to 1 decimal place.
b) The position of the ladder is now changed so that the distance from the foot of the ladder to the base of the wall is the same as the distance the ladder reaches up the wall. How high up the wall does the ladder now reach, to 1 decimal place?

11 The length of the diagonal of a rectangle is 12cm. The length of its base is three times that of its height.
Calculate the length of the base.

Trigonometry 1, 2 & 3

1 Use trigonometry to calculate the length of...

a) BC

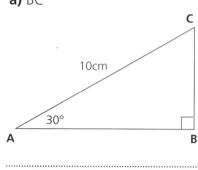

b) AB to 3 significant figures.

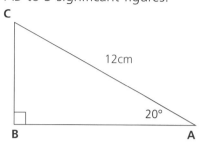

c) AB to 1 decimal place.

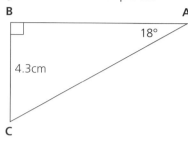

....
....
....
....

....
....
....
....

....
....
....
....

2 Use trigonometry to calculate the size of angle θ in the following triangles. If need be, give your answer to 1 decimal place.

a)

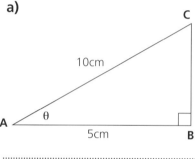

b)

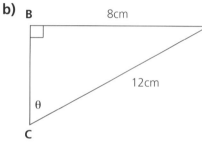

c)

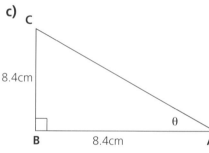

....
....
....
....

....
....
....
....

....
....
....
....

3 Which of the following triangles is right-angled? Show all your workings.
They are not drawn to scale.

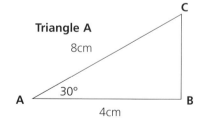

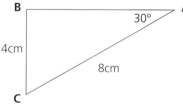

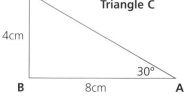

....
....
....
....

....
....
....
....

....
....
....
....

Trigonometry 1, 2 & 3 (cont)

4 The tree in the diagram is 4.2m high. From A the angle of elevation to the top of the tree is 30.4°.

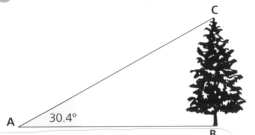

a) Calculate the distance from A to the base of the tree, B, to 1 decimal place.

....

....

....

b) The angle of elevation of the tree is now measured from a position 2m closer to the tree. Calculate the angle of elevation, to 1 decimal place.

....

....

....

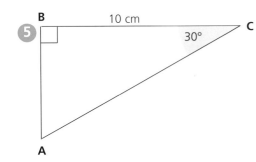

5 The diagram shows a right-angled triangle ABC not drawn to scale. BC = 10cm and angle C = 30°

Angle	Sine	Cosine	Tangent
30°	0.5	0.866	0.577
60°	0.866	0.5	1.73

Without using a calculator, use the table to work out the length of AB.

....

....

....

....

6 Triangle ABC is a right-angled triangle with BÂC = 90°.
Use trigonometry to calculate...
a) Angle ABC to 1 decimal place if AB = 70cm and BC = 1.6m.
b) Angle ABC to 1 decimal place if the length of AC is twice the length of AB.
c) The length of AB to 1 decimal place if AC = 14cm and angle ACB is four times the size of angle ABC.

7 Triangle ABC has AB = 12.6cm, BC = 6.3cm and angle ACB = 60°.
Is triangle ABC a right-angled triangle? Explain your answer.

8 **a)** The angle of elevation of the top of a tower from a point 30m away from the foot of the tower is 48.2°. Calculate the height of the tower to 1 decimal place.
b) The tower has a flag and the angle of elevation of the top of the flag from the same point is 50.4°. Calculate the height of the flag to 1 decimal place.

9 The diagram (not drawn to scale) shows a picture hanging on a wall. AD = 3m, AB = 1.4m and CD = 1.8m.
Calculate the height and width of the picture.

1 **The grid shows twelve shapes A to L.**

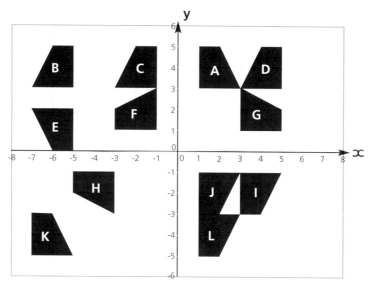

a) Give the letter of the image which is object A reflected...

i) in the x-axis .... **ii)** in the y-axis.. **iii)** in the line $x = 3$ ....

iv) in the line y = 1.............................. **v)** in the line y = x .... **vi)** in the line y = -x ....

b) Describe the single reflection which takes B to E. ....

c) Describe the single reflection which takes F to G. ....

2 **Triangles A, B, C and D are all different reflections of the green triangle. For each one draw in the mirror line and describe the reflection fully.**

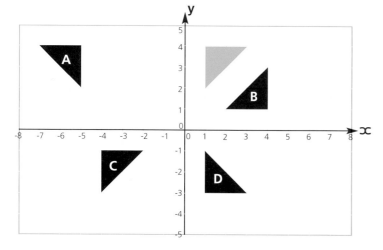

Triangle A: ....

....

Triangle B: ....

....

Triangle C: ....

....

Triangle D: ....

....

3 **A is a triangle with coordinates (3,1), (5,1) and (5,2).**

a) On a suitable grid draw triangle A.

b)i) Reflect triangle A in the line $x = 1$. Label this triangle B. **ii)** Reflect triangle A in the line y = -1. Label this triangle C.

iii) Reflect triangle A in the line y = x. Label this triangle D. **iv)** Reflect triangle A in the line y = -x. Label this triangle E.

Transformations 2

1 The grid shows ten shapes A to J.

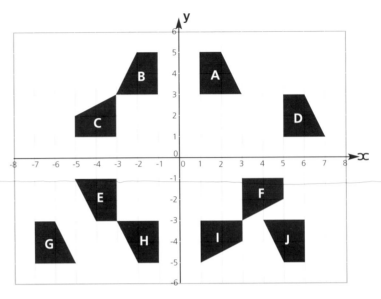

a) Give the letter of the image which is object A rotated…

i) 90° clockwise about the origin (0,0)........................

ii) 180° clockwise about the origin (0,0)...........................

iii) 270° clockwise about the origin (0,0)................

iv) 90° clockwise about the centre (-2,0)........................

v) 180° clockwise about the centre (-1,1)..............

b) Describe the single rotation which takes C to A...

c) Describe the single rotation which takes I to A..

The grid shows triangle A.

2 **a)** Rotate triangle A 90° clockwise about the origin (0,0). Label this triangle B.

b) Rotate triangle A 180° clockwise about the origin (0,0). Label this triangle C.

c) Rotate triangle A 270° clockwise about the origin (0,0). Label this triangle D.

d) Rotate triangle A 90° clockwise about the centre (3,1). Label this triangle E.

e) Rotate triangle A 270° clockwise about the centre (-1,1). Label this triangle F.

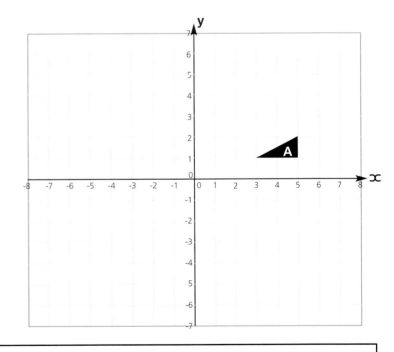

3 **A is a quadrilateral with coordinates (2,2), (4,4), (5,3) and (5,1).**
a) On a suitable grid draw quadrilateral A.
b) i) Rotate quadrilateral A 90° clockwise about the origin (0,0). Label this quadrilateral B. **ii)** Rotate quadrilateral A 180° clockwise about the origin (0,0). Label this quadrilateral C. **iii)** Rotate quadrilateral A 270° clockwise about the origin (0,0). Label this quadrilateral D. **iv)** Rotate quadrilateral A 90° clockwise about the centre (2,2). Label this quadrilateral E.
v) Rotate quadrilateral A 180° clockwise about the centre (-1,-1). Label this quadrilateral F.

1 **The grid shows ten shapes A to J.**

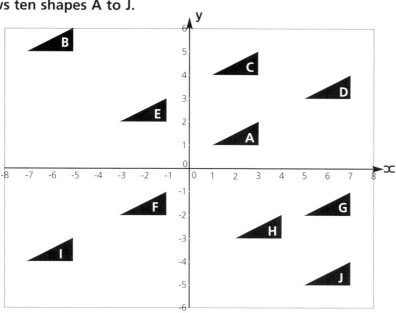

a) Give the letter of the shape which is shape A translated by the vector...

i) $\begin{pmatrix} 4 \\ 2 \end{pmatrix}$ ....

ii) $\begin{pmatrix} 1 \\ -4 \end{pmatrix}$ ....

iii) $\begin{pmatrix} -4 \\ 1 \end{pmatrix}$ ....

iv) $\begin{pmatrix} 0 \\ 3 \end{pmatrix}$ ....

v) $\begin{pmatrix} -8 \\ -5 \end{pmatrix}$ ....

vi) $\begin{pmatrix} 4 \\ -6 \end{pmatrix}$ ....

b) What is the translation vector which maps triangle D onto triangle A?..

c) What is the translation vector which maps triangle B onto triangle A? ....

2 **The grid shows triangle A. Draw and label the following translations.**

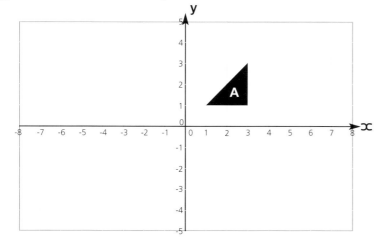

a) Triangle A is mapped onto triangle B by the translation vector $\begin{pmatrix} 4 \\ 2 \end{pmatrix}$

b) Triangle A is mapped onto triangle C by the translation vector $\begin{pmatrix} -5 \\ -4 \end{pmatrix}$

c) Triangle A is mapped onto triangle D by the translation vector $\begin{pmatrix} -8 \\ 0 \end{pmatrix}$

d) Triangle A is mapped onto triangle E by the translation vector $\begin{pmatrix} 5 \\ -6 \end{pmatrix}$

3 **A is a triangle with coordinates (-1,-3), (-3,-2) and (-4,4).**
   a) On a suitable grid draw triangle A.
   b)i) Translate triangle A by the vector $\begin{pmatrix} 5 \\ 5 \end{pmatrix}$ Label this triangle B. **ii)** Translate triangle A by the vector $\begin{pmatrix} -5 \\ -5 \end{pmatrix}$ Label this triangle C.

   iii) Translate triangle A by the vector $\begin{pmatrix} 5 \\ -5 \end{pmatrix}$ Label this triangle D. **iv)** Translate triangle A by the vector $\begin{pmatrix} -5 \\ 5 \end{pmatrix}$ Label this triangle E.

Transformations 4

1 **The grid shows triangle A.**

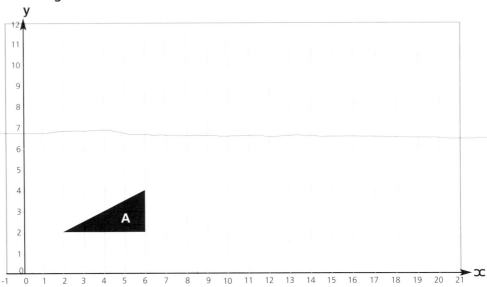

Enlarge triangle A by...

a) a scale factor of 3 about the centre of enlargement (0,0). Label this triangle B.

b) a scale factor of 2 about the centre of enlargement (0,3). Label this triangle C.

c) a scale factor of ½ about the centre of enlargement (0,0). Label this triangle D.

d) a scale factor of 2 about the centre of enlargement (4,0). Label this triangle E.

2 **The grid shows three quadrilaterals A, B and C.**

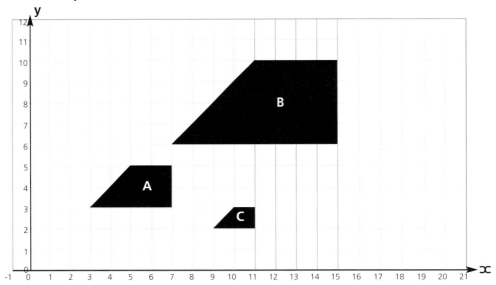

Describe fully the enlargement that would map quadrilateral A...

a) onto B...

b) onto C. ....

---

3 **Triangle A has coordinates (3,2), (6,2) and (6,4).** **a)** On a suitable grid draw triangle A.
b) Complete the following enlargements: **i)** Triangle A is enlarged by a scale factor of 2, centre of enlargement (0,0).
Label this triangle B. **ii)** Triangle A is enlarged by a scale factor of ½, centre of enlargement (0,0). Label this triangle C.
iii) Triangle A is enlarged by a scale factor of 3, centre of enlargement (1,1). Label this triangle D.

1 **The grid shows three triangles A, B and C.**

a) Describe fully a single transformation that would map triangle A onto...

i) triangle B. ....

ii) triangle C. ....

b) Describe fully a single transformation that would map triangle B onto triangle C.

....

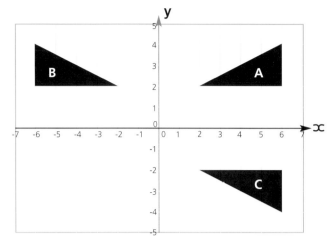

2 **The grid shows three quadrilaterals A, B and C.**

a) Describe fully a single transformation that would map quadrilateral A onto...

i) quadrilateral B ....

ii) quadrilateral C ....

b) Describe fully a single transformation that would map quadrilateral B onto quadrilateral C.

....

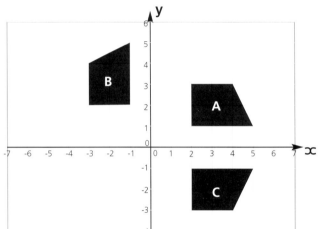

3 **The grid shows triangle A.**

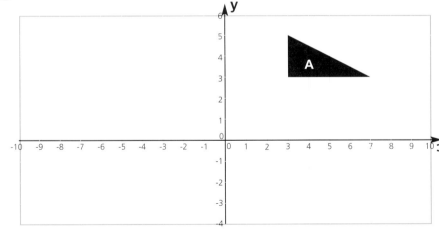

a) Triangle A is reflected in the line $x = 1$. Draw and label this triangle B.

b) Triangle A is reflected in the line y = 1. Draw and label this triangle C.

c) Describe fully the single transformation that would map triangle B onto triangle C.

....

....

4 a) Quadrilateral A has coordinates (3,3), (6,3), (7,6) and (5,6). On a suitable grid draw quadrilateral A.

b) Quadrilateral A is reflected in the line $x = 2$. Draw and label quadrilateral B.

c) Quadrilateral A is reflected in the line y = 0. Draw and label quadrilateral C.

d) Quadrilateral A is rotated 270° clockwise about the origin (0,0). Draw and label quadrilateral D.

e) Describe fully the single transformation that would map quadrilateral B onto quadrilateral C.

f) Describe fully the single transformation that would map quadrilateral D onto quadrilateral C.

Coordinates 1

1 Write down the coordinates of the letters A to H on the graph opposite.

....

....

....

....

....

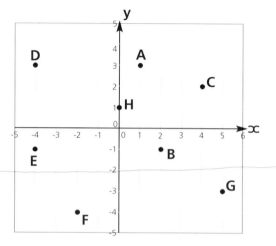

2 **P, Q, and R are 3 corners of a parallelogram.**

a) Write down the coordinates of P, Q and R.

....

....

....

b) Plot the fourth corner of the parallelogram on the axes above labelling it S. Write down the coordinates of S.

....

c) Find the coordinates of the mid points of the line segments:

i) PQ ....

....

ii) QR ....

....

iii) RS ....

....

iv) PS ....

....

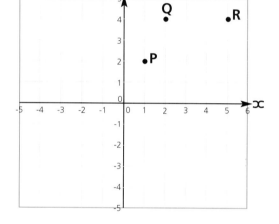

1 a) On the graph opposite, plot the coordinates
A(1,2), B(3,5), C(7,6) and D(7,3).

b) Find the coordinates of the midpoint, M,
of the line segments:

i) AB ....

....

ii) BC ....

....

iii) CD ....

....

iv) AD ....

....

2 a) On the graph opposite, plot the coordinates
P(3,1), Q(0,3), R(4,6), S(7,4).

b) Calculate the lengths of the line segments:

i) PQ ....

....

....

....

ii) QR ....

....

....

....

....

iii) RS ....

....

....

....

iv) SP ....

....

....

....

---

3 The line segment XY has coordinates (-2,-1) and (6,4) respectively. Find the coordinates of the midpoint M.

4 Hexagon ABCDEF has coordinates A(-2,3), B(2,3), C(4,0), D(2,-4), E (-2,-4), F(-4,0).
   a) Plot these on a graph to produce a hexagon.
   b) Find the coordinates of the midpoint, M, for line segments: **i)** AF **ii)** BC **iii)** CD **iv)** EF.
   c) Calculate the lengths of the line segments and hence find the perimeter of the hexagon (to 1 d.p.).

5 Draw axes from -5 to 5 on x-axis and -3 to 4 on y-axis.
   a) Draw and label A (4,2), B (-5,2), C (-4,-2)
   b) Draw and label a fourth coordinate D to produce a parallelogram ABCD.

Perimeter 1 & 2

1 Calculate the perimeter of the following shapes...

a)

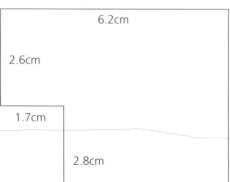

6.2cm

2.6cm

1.7cm

2.8cm

b)

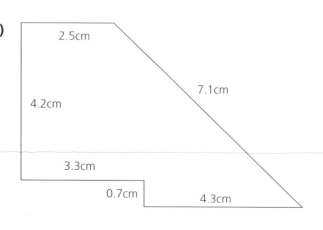

2.5cm

4.2cm

7.1cm

3.3cm

0.7cm

4.3cm

....

....

c)

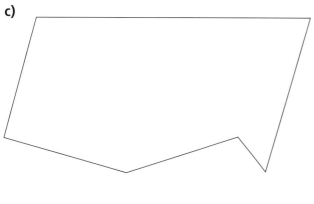

d)

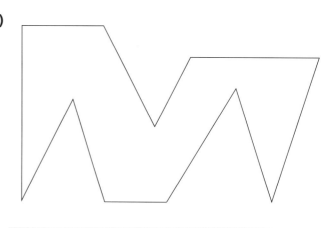

....

....

e)

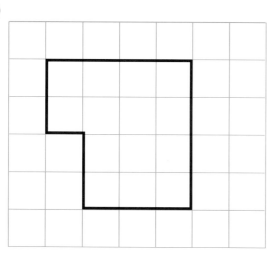

f)

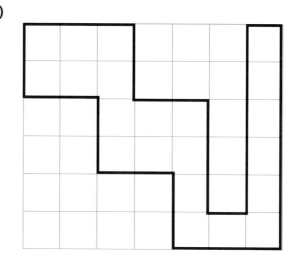

....

....

2 Calculate the circumference of the following circles to 1 decimal place, centre 0. Take π = 3.14.

a)

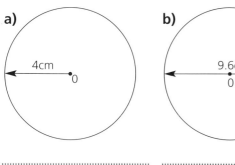

4cm

0

b)

9.6cm

0

....

....

....

....

....

....

....

....

3 A tin has a diameter of 8cm and a height of 15cm. It has a label around it with a 1cm overlap. There is a 1cm gap between the label and the top and bottom of the tin. Calculate...

a) The height of the label.

....

....

b) The length of the label.

....

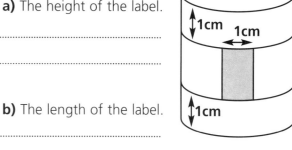

1cm 1cm

1cm

Not to scale

4 A window is in the shape of a semi-circle on top of a rectangle as shown in the diagram.
If AB = 1.8m and BC = 90cm, calculate the perimeter of the window to 1 decimal place.
Take π = 3.14.

D   C

A   B

....

....

....

....

....

....

5 Mrs Jones' garden is rectangular. At each end there is a semi-circular flower bed and the rest of the garden is lawn, as shown in the diagram. If AB = 10m and BC = 8m, calculate the perimeter of the lawn. Take π = 3.14.

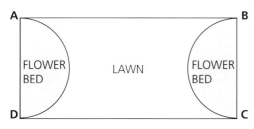

A   B

FLOWER BED   LAWN   FLOWER BED

D   C

....

....

....

....

6 A window measures 2×10^2 m by 1×10^2 m. Calculate the perimeter of the window to 1 decimal place. Give your answer in standard form.

7 A wall measures 3.6×10^3 m by 1.5×10^4 m. Calculate the perimeter of the wall to 1 decimal place. Give your answer in standard form.

8 Calculate the circumference of the following circles to 1 decimal place. Take π = 3.14.
a) Radius = 12cm **b)** Radius = 1.2cm **c)** Diameter = 12cm **d)** Diameter = 120cm

Area 1 & 2

1 Calculate an approximate area for the following shapes. Each square has an area of 1cm².

a)

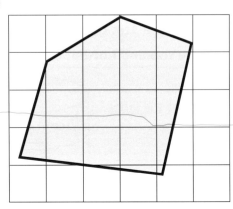

b)

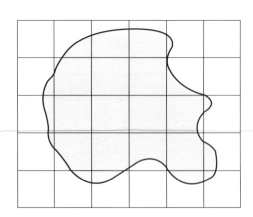

....

....

2 Calculate the area of the following shapes. They are not drawn to scale. Take π = 3.14.

a)

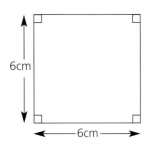

6cm

6cm

b)

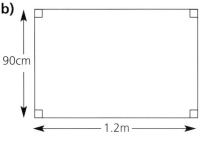

90cm

1.2m

c)

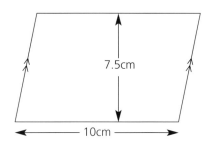

7.5cm

10cm

....

....

....

....

....

....

d)

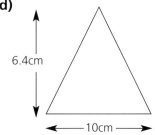

6.4cm

10cm

e)

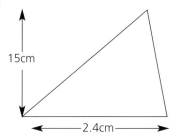

15cm

2.4cm

f)

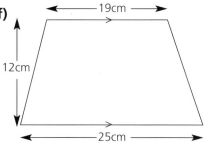

19cm

12cm

25cm

....

....

....

....

....

....

g)

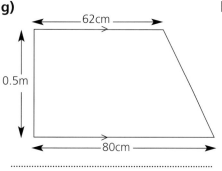

62cm

0.5m

80cm

h)

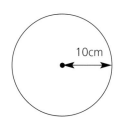

10cm

i)

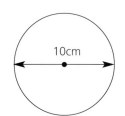

10cm

....

....

....

....

....

....

3 A circular table has an area of 11.304m². Calculate its radius to 1 d.p. Take π = 3.14.

....

....

....

....

4 A circular flower bed has an area of 7.065m². Calculate its radius to 1 d.p. Take π = 3.14.

....

....

....

....

5 Calculate the area of the following shaded shapes to 1 d.p. They are not drawn to scale. Take π = 3.14.

a)

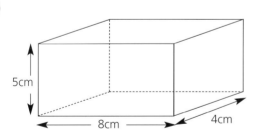

b)

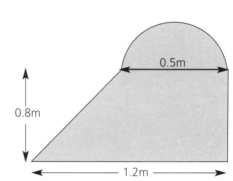

....

....

....

....

....

....

....

....

6 Calculate the surface area of the following solids to 1 d.p.

a)

b)

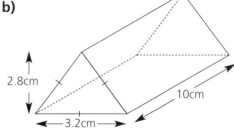

....

....

....

....

7 A property developer decides to varnish the floorboards in the living room of one of his houses. A plan of the floor is shown alongside. It is not to scale. One tin of varnish will cover 16m² and costs £4.99. If he wants to apply two coats of varnish to the whole floor, how much will it cost him?

3m

5.5m

2.5m

8m

Volume 1 & 2

1 Calculate the volume of the following solids.

a)

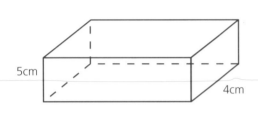

5cm
4cm

b)

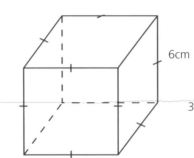
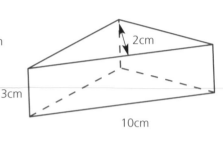

6cm

c)

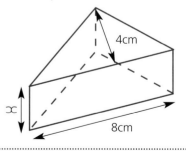

2cm
3cm
10cm

....
....
....
....

2 Calculate the volume of the following solids.

a)

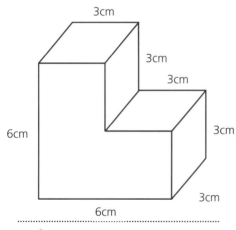

3cm
3cm
3cm
6cm
3cm
3cm
6cm

b)

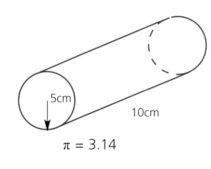

5cm
10cm
3cm

π = 3.14

c)

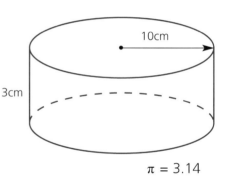

10cm
3cm

π = 3.14

....
....
....
....

3 The following solids all have a volume of 100cm³. For each solid calculate the missing length represented by x.

a)

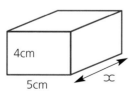

4cm
5cm
x

b)

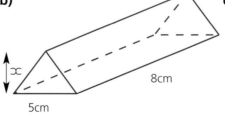

x
5cm
8cm

c)

4cm
x
8cm

....
....
....
....

4 The following cylinders both have a volume of 314cm³. For each cylinder calculate the missing length represented by x (π=3.14).

a)

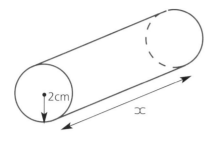

b)

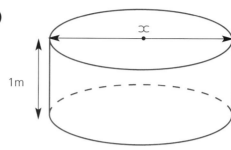

....

....

....

....

5 The tank opposite contains 60 litres of water. All the water is poured into a cylindrical tank which has a diameter of 44cm. Calculate the depth of the water in the cylindrical tank (to 1d.p.) π = 3.14.

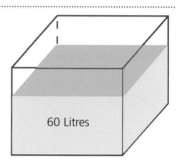

60 Litres

....

....

....

....

6 In the following expressions the letters r and h represent length. π, 2, 3, 4, 5, and 10 are numbers which have no dimension. Write down whether each of the following expressions represent perimeter, area, volume or none of them.

a) $r(\pi + 2)$ ....

b) $\dfrac{4r^2\pi}{h}$ ....

c) $r(r + 4h)$ ....

d) $\dfrac{rh}{4}$ ....

e) $10r^3\pi$ ....

f) $\pi(r+2h)$ ....

g) $\dfrac{3r^3}{h}$ ....

h) $r^2(h + \pi r)$ ....

---

7 A cylindrical mug has internal radius 5cm and internal height 8cm.
   a) Calculate the volume of liquid it can hold (to 3 sig. fig.) (π = 3.14)
   b) If 500cm³ of liquid is poured into the mug, what would the depth of liquid be in the mug?

8 If x and y both represent length, which of the following expressions represent i) length, ii) area, iii) volume?
   $\dfrac{x}{y}$, $\sqrt{x^3}$, $x^2 + y^2$, $\dfrac{x^2}{y}$, xy^2, x^2y^2, $x^2 + y^3$

3-D Shapes 1 & 2

1 a) What shape is the base of the
cuboid shown opposite? ....

b) Which edges are equal in
length to CD?...

c) Which lengths equal AH? ....

d) How many vertices has the cuboid? ....

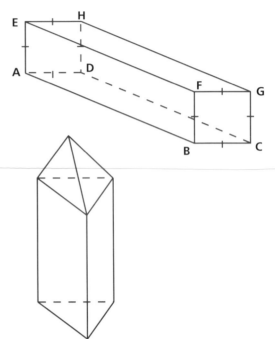

2 **An equilateral triangular prism has a tetrahedron**
placed on top of it. For this combined solid...

a) How many edges does it have? ....

b) How many vertices? ....

c) How many faces? ....

3 **On the grid below draw full size diagrams of the following solids.**

a)

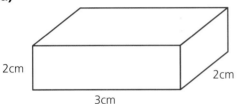

2cm
3cm
2cm

b)

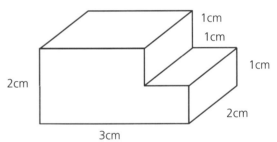

1cm
1cm
1cm
2cm
2cm
3cm

4 **Draw a sketch of the plan view of this**
square based pyramid.

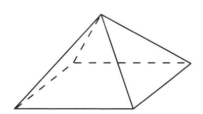

3-D Shapes 1 & 2 (cont)

5 The diagram alongside shows a solid. Draw and label an accurate diagram of the solid showing...

a) plan view **b)** front elevation **c)** side elevation

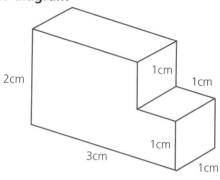

6 A net of a solid is shown opposite.

a) What is the name of the 3-D solid?

....

b) How many vertices does it have?

....

c) Which other corners meet at D? Put an X on each one.

d) How many planes of symmetry does the solid have?

....

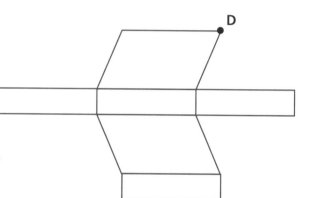

7 Which of the following are nets for a triangular prism? Place a tick beside the correct net(s).

a)

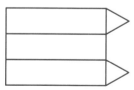

b)

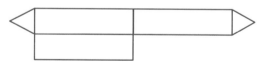

c)

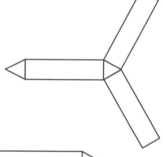

d)

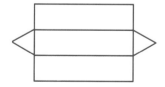

---

8 Draw an accurate full-size net of a regular hexagonal prism if each edge is 3cm long.

9 How many planes of symmetry does a cube have?

10 Draw the **a)** plan **b)** front elevation **c)** side elevation of this solid.

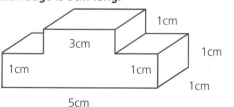

Symmetry 1 & 2

1 Draw all the lines of symmetry for each of the following shapes.

2 Mark on the lines of symmetry for each of these letters.

M S H B Z A V C K

3 How many lines of symmetry does each of the following quadrilaterals have?

a) Square .... ☐   d) Rhombus ....

b) Rectangle .... ▭   e) Kite ....

c) Parallelogram .... ▱   f) Trapezium ....

4 For each of the following shapes draw in all the lines of symmetry and also write down the order of rotational symmetry for each shape.

a)

Order of rotational symmetry:

....

b)

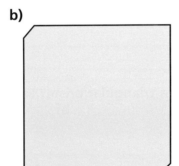

Order of rotational symmetry:

....

c)

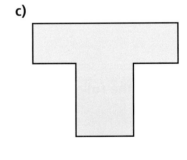

Order of rotational symmetry:

....

5 What is the order of rotational symmetry for each of these shapes?

a)

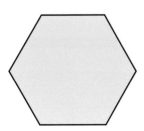

Order: ....

b)

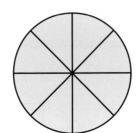

Order: ....

c)

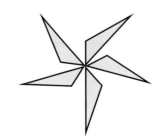

Order: ....

d)

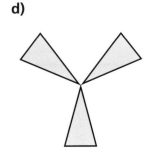

Order: ....

6 **a)** A pattern has four lines of symmetry. Part of the pattern is shown. Complete the pattern.

b) A different pattern has rotational symmetry order 4 and no line symmetry. Part of the pattern is drawn. Complete the pattern.

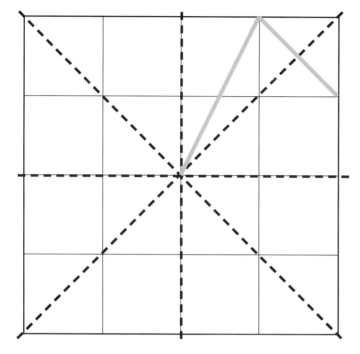
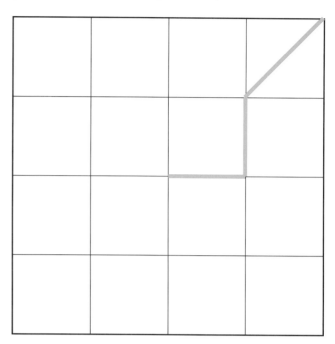

7 **a)** Draw accurately a regular pentagon with sides 3cm long and internal angles 108°.

b) Mark on the lines of symmetry.

c) What is the order of rotational symmetry?

8 **a)** Draw a regular hexagon. Interior angles are 120°.
b) Draw in all lines of symmetry and write down the order of rotational symmetry.

9 **Write down the order of rotational symmetry for each quadrilateral:**
a) Square **b)** Rectangle **c)** Parallelogram **d)** Rhombus **e)** Kite **f)** Trapezium.

Scale Drawings & Map Scales

1 a) Draw an accurate scale drawing of this garden, using a scale of 1cm to represent 2.5m.

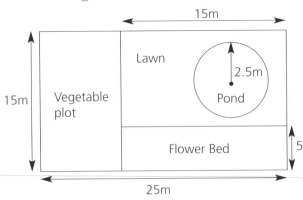

b) What is the actual diagonal distance across the garden in metres?

....

2 This is part of a map of Devon and Cornwall drawn to a scale of 1cm : 10km.

a) What is the direct distance from Launceston to Exeter?

....

....

b) What is the direct distance between Bodmin and Looe?

....

....

c) Which two places are a direct distance of 43km from Tavistock?

....

....

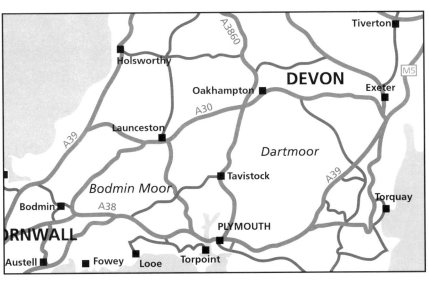

3 Draw an accurate scale drawing of a rectangular field 80m long and 50m wide. By measurement, find the actual distance diagonally from one corner to the opposite corner, to the nearest metre.

4 The diagram alongside shows a sketch of one side of a house.
   a) Draw an accurate scale drawing using a scale of 1cm to 1m.
   b) By measurement, find the actual height of the house (x).

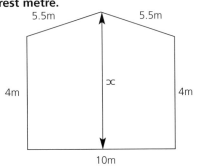

Measuring Bearings

1 The diagram shows the position of the coastguard (C), the beach (B) and a yacht (Y).

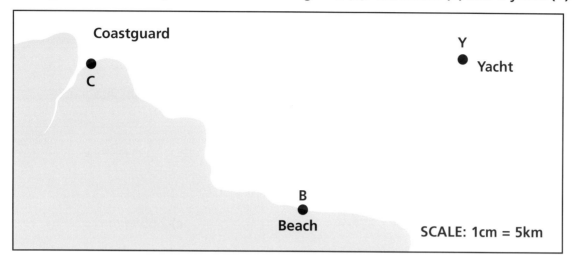

Coastguard

Y
● Yacht

● C

B
●
Beach

SCALE: 1cm = 5km

a) What is the bearing of...

i) Y from C? ....

ii) Y from B? ....

iii) B from C? ....

iv) B from Y? ....

v) C from Y? ....

vi) C from B? ....

b) What is the actual distance from...

i) C to B? ....

ii) C to Y? ....

iii) B to Y? ....

2 The map shows the position of 4 towns A, B, C and D on an island.

A helicopter flies directly from A to B, then B to C, then C to D and finally D back to A.

On what four bearings must it fly?

A➤B ....

B➤C ....

C➤D ....

D➤A ....

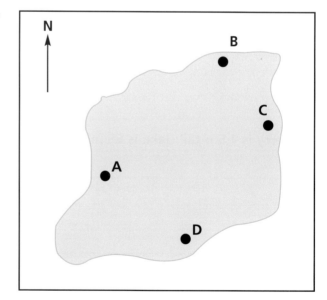

3 An explorer walks 1 000m on a bearing of 070° and then walks 2 000m on a bearing of 160°.

a) Draw an accurate scale drawing of his route.

b) By measurement, find the bearing he must follow to return directly to his starting point.

4 Treasure is buried on an island according to the following instructions: "The treasure lies on a bearing of 100° from the coconut tree and on a bearing of 200° from the cactus plant. The cactus plant is 20m due east of the coconut tree." Draw a scale drawing using 1cm to 5m to show the position of the treasure.

Converting Measurements

1 **Convert...**

a) 450cm into metres

....

....

b) 3.5 litres into millilitres

....

....

c) 1.25kg into grams

....

....

d) 6 874g into kg

....

....

e) 45km into metres

....

....

f) 0.55cm into mm

....

....

2 **Convert these lengths into metres.**

a) 1 005cm

....

....

b) 1.937km

....

....

c) 2 650mm

....

....

3 **Put these weights into order of size, smallest first.**

420g, 4kg, 39.5kg, 4 220mg, 0.405kg

....

....

....

4 **Convert...**

a) 45cm into inches

....

....

b) 6 ounces into grams

....

....

c) 5 gallons into litres

....

....

5 **Terry is 1.5m tall. Jake is 68 inches tall. Who is taller and by how much?**

....

....

....

6 **Put these lengths into decreasing order of size:**
1km, 900m, 1 200m, 11 000cm, 1 050 000mm

7 **Convert...**
a) 3 miles into km **b)** 12km into miles **c)** 4.5 pounds into grams **d)** 360g into pounds **e)** 12 pints into litres
f) 22.5 litres into pints

8 **Sue ran a 10km race. How many yards did she run altogether? (1760 yards = 1 mile).**

Compound Measures

1. Work out the time taken to travel 92km at an average speed of 55km/h.

   ....

   ....

2. A marathon runner completes 26.2 miles in a course record of 2hrs 20mins.
   What was his average speed?

   ....

   ....

3. A tortoise takes 20 minutes to get from one end of the garden to the other. His average speed
   was 2cm per second. How long is the garden in metres?

   ....

   ....

   ....

4. A boat travels for $2\frac{1}{2}$ hours at 100km/h and then $1\frac{1}{2}$ hours at 80km/h. Calculate its average speed
   for the whole journey.

   ....

   ....

   ....

5. A cylindrical tree stump weighs 15kg.
   Its dimensions are shown opposite.
   Calculate the density of the wood.

   ....

   ....

   ....

   50cm

   30cm

6. Water has a density of 1g/cm^3 and ice has a density of 0.9g/cm^3.
   450cm^3 of water is frozen. By how much does the volume of the water change when it freezes?

   ....

   ....

   ....

7. **a)** Change 30 metres per second into km per hour. **b)** A car travels 30 metres per second for $3\frac{1}{2}$ hours. How far does it
   travel in km?

8. **The density of oak wood is 800kg per m^3.**
   a) Change this to g per cm^3.
   b) How much does a solid oak table top measuring 110cm by 80cm by 5cm weigh?

Constructions 1 & 2

1 Construct a triangle ABC with sides AB = 4cm, BC = 3cm and angle ABC = 40°.

2 On your triangle ABC in question 1 above, construct the bisector of angle BAC.

3 **a)** Construct a triangle PQR where PQ is 6cm, QR is 5cm and PR is 2cm.

   b) Construct the perpendicular bisector of PQ and where this line meets QR, label it S.

4 The sketch opposite shows three towns A, B and C. B is on a bearing of 040° from A. C is due East of A and B is on a bearing of 300° from C. C is 40km from A. Construct an accurate scale drawing.

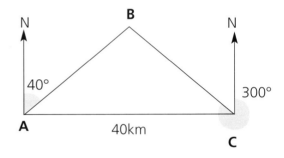

5 Showing all construction lines, and by using a pair of compasses and a ruler, construct the perpendicular bisector of each side of this equilateral triangle.

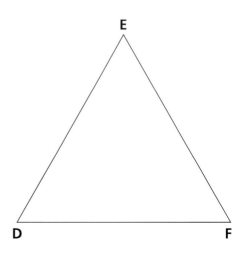

6 **Using the line AB below as a starting point, use a pair of compasses and a ruler to...**

a) Construct a 90° angle at A

b) Construct a 90° angle at B

c) Complete the construction to make a square.

A   **B**

7 **In the space below draw a triangle XYZ.**

a) Construct the bisectors of the three angles $\hat{X}$, $\hat{Y}$ and $\hat{Z}$.

b) What do you notice? ....

8 Using only a ruler and a pair of compasses, construct an equilateral triangle of side 3cm and a square of side 7cm.

9 Construct the following triangles using a ruler and a pair of compasses.

a)

7.4cm   5.3cm

5.3cm

b)

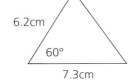

6.2cm

60°

7.3cm

c)

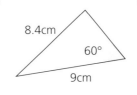

8.4cm

60°

9cm

d)

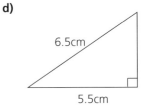

6.5cm

5.5cm

Loci

1 The map below shows three trees, A, B and C, in a park. Along one edge of the park there is a straight path. Treasure is buried in the park. The treasure is: nearer to C than A, more than 150m from the path and between 100m and 200m from B. Using a ruler and compass only, shade the region where the treasure must be buried. You must show all construction lines.

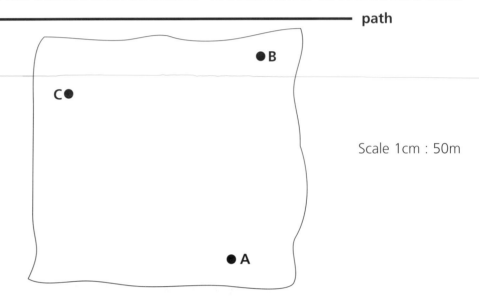

Scale 1cm : 50m

2 A goat is tethered by a rope 4m long to a rail PQ 8m long. The rope can move along the rail from P to Q. Draw an accurate diagram of the locus of points showing the boundaries of the area where the goat can graze.

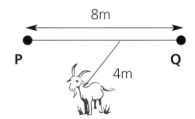

3 Draw a triangle ABC where AB = 7cm, BC = 9cm, $\hat{B}$ = 90°.
a) Draw the locus of points inside the triangle which are equidistant from A and B.
b) Draw the locus of points inside the triangle which are equidistant from B and C.
c) Find and label a point D which is equidistant from A and B and equidistant from B and C.

4 Draw the same triangle ABC as for question 3.
a) Draw the locus of points equidistant from line AB and AC.
b) A point (P) moves inside triangle ABC, equidistant from AB and AC and greater than 6cm away from C. Show the locus of points where P can move.

5 **a)** Draw the graph of the set of points which are equidistant from the x and y axes.
b) Write down the equation of the graph.
c) On the same axes draw the graph of the set of points where the y coordinate is twice the x coordinate.
d) What is the equation of this graph?

Properties of Circles 1 & 2

1 **Name each of the parts of the circle labelled A to F opposite.**

a) A .... b) B ....

c) C .... d) D ....

e) E .... f) F ....

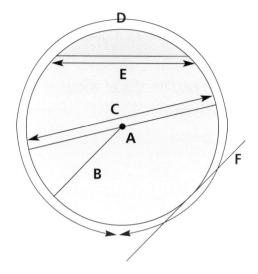

2 **A circle has two points A and B on its circumference, so that the chord AB cuts the circle into two segments.**

a) Draw a sketch of the circle and the chord AB.

b) Label the two segments 'major' and 'minor'.

3 **In the diagram opposite, O is the centre of the circle. AB is a tangent touching the circle at C. OC = AC. Find the size of...**

a) angle COA

....

....

b) angle ODC

....

....

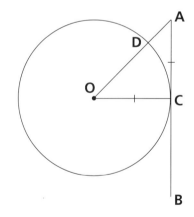

4 **XT and XW are tangents touching the circle at T and W.**

a) Draw the axis of symmetry in OTXW.

b) Name three pairs of congruent triangles.

....

....

....

c) If angle TXW is 40° what is angle TOW?

....

....

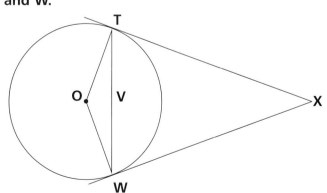

5 In the diagram opposite, O is the centre of the circle and PQ and PR are tangents.

a) What name is given to triangle OQR?

....

b) Find the size of angle...

i) OQR ....

ii) ORQ ....

c) Find the size of angle...

i) RQP..

ii) QRP ....

iii) QPR ....

d) What type of triangle is PQR?

....

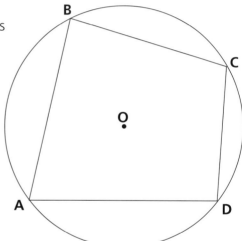

6 a) On the circle opposite, use a ruler and a pair of compasses

to construct the perpendicular from the centre, O, to each

of the four chords AB, BC, CD and AD.

b) The perpendicular to AD touches AD at X.

Using triangle OAD explain why AX = XD.

....

....

....

....

7 A is a point on the circumference of a circle

with centre O. AB is a tangent length 6cm.

The area of triangle OAB is 30cm². Without

using a calculator work out the area of the

circle.   Give your answer in terms of π.

....

....

....

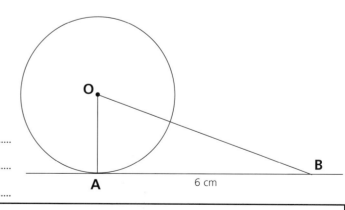

8 **a) i)** In diagram A, OX̂Y= 90°.
Explain why OX bisects WY.

ii) Calculate the length OX if
the chord is 10m long and
radius OW is 8m long.

Diagram A

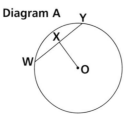

b) In diagram B, AB and CD
are parallel chords. OPQ is
perpendicular to AB and CD.
Explain why AC = BD.

Diagram B

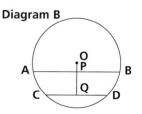

Properties of Circles 3

1 Find the size of x. Give a reason for your answer.

a)
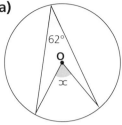

b)
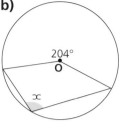

c)
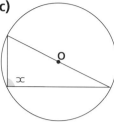

....
....

2 Find the size of x and y. Give a reason for your answer.

a)
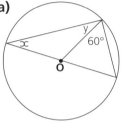

b)
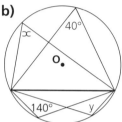

c)
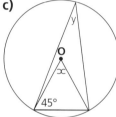

....
....

3 Find the size of angle AED. Give a reason for your answer.

a)
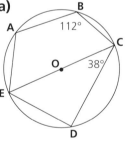

b)
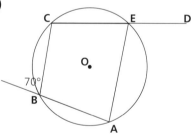

....
....

4 Calculate the value of x.

a)
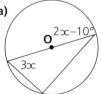

b)
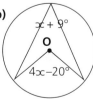

c)
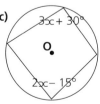

5 Which one of the following kites is a cyclic quadrilateral. Give a reason for your answer.

Kite 1   Kite 2   Kite 3

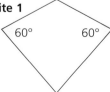

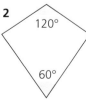

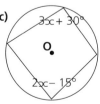

Probability

1 Tim has ten cards (shown below). They are placed face down and mixed up.

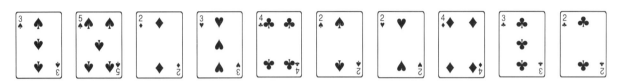

a) What is the probability that a card picked at random will be...

i) a 2? ....

ii) a 3? ....

iii) a 4? ....

iv) a 5? ....

v) not a 2? ....

vi) not a 3? ....

b) Here is a probability scale:

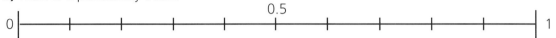

On the scale above ....

i) mark with an A the probability that a card picked at random will be a 2.

ii) mark with a B the probability that a card picked at random will be a 3.

iii) mark with a C the probability that a card picked at random will be a 4.

iv) mark with a D the probability that a card picked at random will be a 5.

2 Fifty people take a driving test at centre A on one day. The table shows the results. A person is chosen at random from the group.

	Pass	Fail
Male	13	15
Female	9	13

a) What is the probability that the person is male?

....

b) What is the probability that the person passed the test?

....

c) In Britain the probability of a person passing their driving test is 0.7. Ben says it is easier to pass at centre A. Explain why Ben is wrong.

....

3 A page in a calendar shows the following month of June.

If a date is chosen at random. What is the probability that it is a ...

a) an even number? ....

b) an odd number? ....

c) a prime number? ....

d) a weekday? ....

e) not a weekday? ....

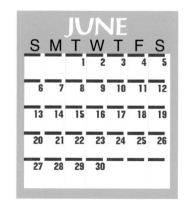

Mutually Exclusive Outcomes

4 **Vicky has two sets of cards, A and B as shown below:**

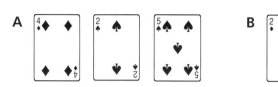

a) Complete all the possible outcomes if two cards are picked, one from A and one from B, at random.

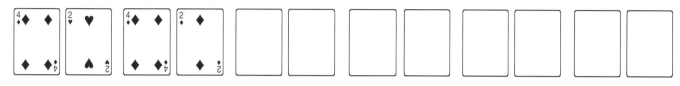

b) What is the probability that the two cards picked are **i)** both 2s? ....

ii) both the same suit? .... **iii)** not the same suit? ....

5 **On her way to work, Safia has to drive through a set of traffic lights. The probability that Safia will arrive at work on time is 0.4 and that she will arrive late is 0.25. What is the probability that she will arrive early?**

....

6 **A bag contains 8 blue marbles, 6 red marbles and 2 green marbles. One marble is pulled out at random and not replaced back into the bag. A second marble is then picked out at random.**

a) What is the probability that the second marble picked out is blue, if the first marble picked out was green? ....

b) What is the probability that the second marble picked out is green, if the first marble picked out was also green? ....

---

7 **For each of these statements say whether you agree or disagree and give a reason for your answer.**
a) The probability it will rain tomorrow is $\frac{4}{3}$.
b) In my tin of sweets there are 12 boiled sweets and 18 chewy sweets, so the probability of choosing a chewy sweet is $\frac{18}{30}$.
c) A fair coin is tossed twice and lands on heads both times. The probability of getting a head the next time the coin is tossed is $\frac{1}{8}$.

8 **A box contains 7 red balls and 3 blue balls. A ball is picked out at random.**
What is the probability that it is **a)** red? **b)** not red? **c)** blue? **d)** not blue? **e)** red or blue? **f)** black?

9 **A kitchen cupboard contains tins of baked beans, peas, carrots and potatoes only. The probability of picking a tin of potatoes is $\frac{1}{12}$, while the probability of picking a tin of carrots is three times that of a tin of potatoes, and a tin of peas is twice that of a tin of carrots.**
a) What is the probability of picking **i)** a tin of carrots? **ii)** a tin of peas? **iii)** a tin of baked beans?
b) What is the probability of not picking **i)** a tin of carrots? **ii)** a tin of peas? **iii)** a tin of baked beans?
c) If the cupboard contains 4 tins of baked beans, how many tins are there in the cupboard altogether?

Listing all Outcomes

1 **Two dice are thrown. The two numbers are added together to give a total score.**

a) Complete the sample space diagram below to show all the scores:

	First Die					
	1	**2**	**3**	**4**	**5**	**6**
1	2	3	4			
2	3	4				
3	4					
4						
5						
6						

Second Die

b) What is the probability that the total score will be ...

i) equal to 7? ....

ii) greater than 7? ....

iii) a prime number? ....

iv) a square number? ....

v) greater than 12? ....

vi) a multiple of 3? ....

vii) a factor of 12? ....

2 **Francis has the following coins in her pocket:**

Jim has the following coins in his pocket:

Two coins are picked out at random, one from Francis' pocket and one from Jim's pocket.

The values of the two coins are added together to give a total.

a) Complete the sample space diagram below to show all the values:

	Francis' Coin					

Jim's Coin

b) What is the probability that the total value of the two coins added together is ...

i) equal to 6p? ....

ii) equal to 11p? ....

iii) less than 10p? ....

iv) greater than 40p? ....

v) less than 40p? ....

vi) equal to 40p? ....

3 **Bruce has an ordinary die. Robin has the following spinner:**
The spinner is spun and the die is thrown to give two numbers.
a) Draw a sample space diagram to show all the possible scores if the numbers are multiplied together.
b) What is the probability that the score is **i)** equal to 12? **ii)** equal to 24? **iii)** a multiple of 10?
iv) a factor of 4? **v)** an odd number? **vi)** an even number?

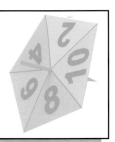

1 **The following spinner is spun 120 times:**

a) How many times would you expect the spinner to land on ...

i) a 1? .... **ii)** a 2? .... **iii)** a 3? ....

b) The actual number of times the spinner landed on a 1, 2 and 3 is shown in the table below. For each number calculate the relative frequency.

Number	Number of times landed	Relative Frequency
1	66	....
2	38	....
3	16	....

2 **Jane tosses a coin 50, 100, 150, 200 and 250 times. She records the number of tails she gets in a table (shown alongside):**

a) Complete the table by calculating the missing relative frequencies.

b) On the grid below draw a bar graph to show the relative frequency of the coin landing on tails.

Number of tosses	Number of tails	Relative Frequency
50	20	0.4
100	44	....
150	80	....
200	92	....
250	120	....

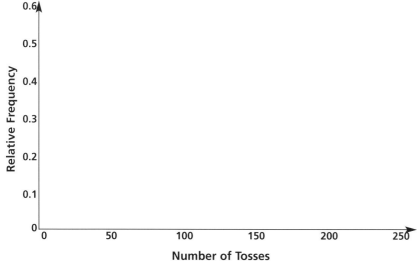

Relative Frequency

Number of Tosses

c) If Jane kept tossing the coin, what would you expect the relative frequency of the coin landing on tails to become? Explain why.

....

....

....

....

....

....

3 **Below are the results of an experiment where a die was thrown and the number of 6s were recorded.**

Number of throws	30	60	90	120	150	180	210	240	270	300	330	360
Number of 6s thrown	3	5	10	16	23	28	35	42	47	49	52	59
Relative frequency												

a) Complete the table by calculating the missing relative frequencies.

b) Draw a bar graph to show the relative frequency of throwing a 6.

c) How many 6s would you expect to be thrown if the experiment was continued and the die was thrown 1500 times?

Tree Diagrams 1 & 2

1 **The probability that Steve arrives at school on time on any particular day is 0.7.**

a) Complete the tree diagram for two school days, Monday and Tuesday.

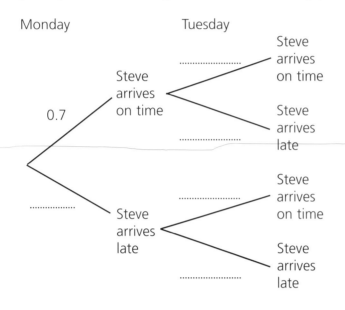

Monday   Tuesday

....
                        Steve
Steve   arrives
arrives   on time
0.7   on time
                        Steve
                        arrives
.... late

                        Steve
.... arrives
Steve   on time
.... arrives
late
                        Steve
                        arrives
.... late

b) What is the probability that …

i) Steve arrives on time on Monday and Tuesday?

....

ii) Steve arrives late on Monday and Tuesday?

....

iii) Steve arrives late on Monday only?

....

iv) Steve arrives late on Tuesday only?

....

v) Steve arrives late on one day only?

....

2 **Vicky, Donna and Petra are going to have two races. The probability that Vicky wins either of the two races is 0.5, while the probability that Donna wins either of the two races is 0.3.**

a) What is the probability of Petra winning either of the two races? ....

b) Complete the tree diagram.

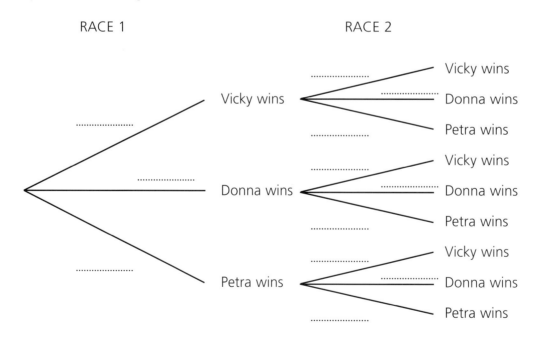

RACE 1   RACE 2

                                        .... Vicky wins
                        Vicky wins   .... Donna wins
.... Petra wins

                                        .... Vicky wins
.... Donna wins   .... Donna wins
                                        .... Petra wins

                                        .... Vicky wins
.... Petra wins   .... Donna wins
                                        .... Petra wins

c) What is the probability that…

i) Vicky wins both races? .... **ii)** Vicky does not win a race? ....

iii) Donna wins Race 1 and Petra wins Race 2? ....

iv) Race 1 and Race 2 are won by different girls? ....

3 **A bag contains 12 counters of which 6 are black, 4 are white and 2 are red. A counter is picked out at random and, without it being replaced, another counter is picked out at random.**

a) In the space below draw a tree diagram to show all the different possible outcomes.

b) What is the probability that...

i) Both counters picked out are the same colour? ....

ii) Both counters picked out are different colours? ....

iii) Neither of the counters picked out are black? ....

---

4 **The probability that Julie does her school homework on any particular day is 0.8.**
a) Draw a tree diagram to show all the probabilities for two days. **b)** Using the tree diagram, what is the probability that Julie **i)** does her homework on both days **ii)** does her homework on one day only and **iii)** does not do her homework on either day?

5 **Rolf is a cricketer. The probability that Rolf's team win a game is $\frac{1}{2}$ and lose is $\frac{1}{3}$. They can also draw. Rolf's team have a cricket game on Saturday and another game on Sunday..**
a) Draw a tree diagram to show all the probabilities. **b)** Using the tree diagram, what is the probability that Rolf's team **i)** win both games **ii)** win on Saturday only **iii)** win one game only **iv)** do not lose either game?

6 **A bag contains 12 balls of which 4 are red, 3 are blue, 3 are green and 2 are yellow. A ball is picked at random and, without it being replaced, another ball is picked at random.**
a) Draw a tree diagram to show all the different possible outcomes.
b) What is the probability that **i)** both balls are the same colour **ii)** both balls are different colours **iii)** neither of the balls picked out are red?

Collecting Data 1 & 2

1 **What is the difference between primary data and secondary data?**

....

....

....

2 **a)** What is sampling?

....

b) A recent survey carried out in Manchester suggests that 90% of the national population prefer football to rugby. The survey was conducted on 200 males. Has this survey provided reliable data? Explain why.

....

....

....

....

....

3 **Jenny is concerned about the breakfast habits of the pupils in her school. She decides to use a questionnaire to find out what pupils have to eat for breakfast. She has chosen eight questions for her questionnaire. Decide whether each question is suitable or not suitable, giving a reason for your answer.**

a) Everybody should eat breakfast. Don't you agree?

....

b) Do you eat breakfast?

....

c) What time do you get up in the morning?

....

d) Are you a vegetarian?

....

e) What year are you in?

....

f) What do you have to eat if you have breakfast?

....

g) If you eat breakfast do you have cereal, toast or other?

....

h) Do you brush your teeth before breakfast?

....

Collecting Data 1 & 2 (cont)

4 Joe works in a supermarket. He decides to use a questionnaire to find out about the shopping habits of the customers who come into the store. Here is the first question:

'How old are you?' Tick the correct box

10 years and under		11 years to 20 years		21 years to 40 years		41 years to 60 years		61 years to 80 years		Over 80 years	

a) Make up two more suitable questions Joe could use for his questionnaire. Include a response section.

....

....

b) Make up two questions which would not be suitable for his questionnaire.

....

....

5 In a survey a group of pupils were asked, 'How long did you spend watching TV over the weekend?'

a) Design a suitable observation sheet to collect this information.

b) How would you make sure that the information obtained was from a random sample?

....

....

6 Molly works in a pizza parlour. She decides to use a questionnaire to find out about the eating habits of the people who come into the parlour.
a) Make up three suitable questions Molly could use for her questionnaire. Include at least one response section.
b) Make up three questions which are not suitable for her questionnaire.

7 Jimmy is carrying out a survey to investigate what the pupils in his school spend their pocket money on.
a) Design a suitable observation sheet for him to collect the information.
b) How could Jimmy make sure that the information collected was random.

Sorting Data 1 & 2

1 What is the difference between discrete and continuous data?

....

....

....

....

2 Jane is carrying out a traffic survey. She records the number of cars that pass her house every 30 seconds for a period of 20 minutes. Group her data by completing the frequency table below.

Number of cars

4	3	4	X	2	5	5	4	3	2
4	5	3	2	4	4	4	5	5	X
2	4	3	5	4	3	3	5	2	4
2	4	3	5	2	X	X	4	5	3

Number of cars	Tally	Frequency
1	IIII	4
2		
3		
4		
5		

3 A survey was carried out on the number of residents in each house on a street.

The results are given below:

4 , 5 , 3 , 4 , 3 , 6 , 1 , 5 , 4 , 2 , 3 , 4 , 5 , 2 , 4 , 5 , 5 , 3 , 4 , 6
4 , 2 , 5 , 3 , 5 , 2 , 4 , 3 , 4 , 1 , 3 , 4 , 4 , 1 , 5 , 2 , 4 , 5 , 3 , 4

a) In the space below group together the results in a frequency table.

b) What percentage of the houses have 3 or more residents? ....

4 **The test results for a group of students are given below.**

Group the data to complete the frequency table below.

31	61	40	63	65
78	52	57	15	35
77	11	68	46	68
64	70	26	87	49

Test Mark	Tally	Frequency
0-19	II	2
20-39		
40-59		
60-79		
80-99		

5 **John has recorded the temperature at midday every day for the month of June using a thermometer. All the temperatures are to the nearest degree Celsius.**

a) Group together John's results by completing the frequency table.

JUNE						
		1 7°C	2 11°C	3 11°C	4 13°C	5 16°C
6 17°C	7 16°C	8 15°C	9 13°C	10 13°C	11 16°C	12 14°C
13 8°C	14 9°C	15 15°C	16 14°C	17 14°C	18 18°C	19 17°C
20 16°C	21 16°C	22 11°C	23 12°C	24 14°C	25 13°C	26 17°C
27 20°C	28 22°C	29 19°C	30 21°C			

Temperature (°C)	Tally	Frequency
$5 \leqslant T < 10$		
$10 \leqslant T < 15$		
$15 \leqslant T < 20$		
$20 \leqslant T < 25$		

b) What percentage of the recorded midday temperatures in June are 15°C or more?

....

6 Bolton Wanderers scored the following number of goals in Premier League matches for the 2002/2003 season.
1, 1, 1, 1, 2, 1, 1, 0, 1, 1, 1, 1, 4, 1, 0, 1, 0, 1, 4
0, 0, 0, 1, 0, 1, 4, 1, 1, 0, 2, 1, 2, 0, 1, 0, 2, 0, 2
Group together the data in a frequency table.

7 Emma decided to measure the height (h) of all the students in her class.
Here are the results, to the nearest cm:
171, 178, 166, 173, 180, 173, 186, 176, 170, 184, 178, 174, 169, 189, 175, 182, 181, 171, 179, 164, 178, 175, 174, 191, 169, 178, 173, 188, 167, 192.

a) Sort the data into a frequency table using class intervals $160 \leqslant h < 165$, $165 \leqslant h < 170$ etc.
b) What percentage of the students in Emma's class have a height measurement of 170cm or more?

8 The individual weights of 40 people, to the nearest kg, are as follows:
79, 75, 68, 70, 83, 72, 81, 89, 61, 74, 80, 51, 84, 63, 73, 54, 76, 74, 80, 85, 94, 77, 71, 81, 70, 66, 87, 62, 59, 63, 63, 67, 75, 81, 80, 78, 60, 77, 61, 75.
Sort the data into a frequency table using appropriate class intervals.

Sorting Data 3

1 a) The following data shows the age (in years) of 30 shoppers in a supermarket.

**41 , 51 , 8 , 60 , 21 , 31 , 41 , 17 , 68 , 28 , 34 , 45 , 46 , 52 , 74 ,
56 , 10 , 23 , 47 , 30 , 34 , 9 , 42 , 29 , 55 , 44 , 38 , 57 , 47 , 58**

Using tens to form the 'stem' and units to form the 'leaves', draw a stem and leaf diagram to show the data.

b) What is the median age of the shoppers? ....

c) What is the range of the ages? ....

**2 A survey of 120 people was conducted to find out if they listened to the radio whilst driving.
Complete the two-way table to show the results.**

	Men	Women	Total
Listen to radio	32		73
Do not listen to radio		24	
Total	55		

3 a) A survey of 200 Year 7, 8 and 9 pupils was carried out to find their favourite type of music from a choice of three: Pop, Rap or Dance. Complete the two-way table to show the results.

	Year 7	Year 8	Year 9	Total
Pop	42		18	
Rap		12		41
Dance	14		31	69
Total			62	

b) What percentage of the pupils chose Pop as their favourite type of music?

....

4 Here are the heights, to the nearest cm, of 30 students in a class:
171, 178, 166, 173, 180, 173, 186, 176, 170, 184, 178, 174, 169, 189, 175, 182, 181, 171, 179, 164, 178, 175, 174, 191, 169, 178, 173, 188, 167, 192.
a) Using tens to form the 'stem' and units to form the 'leaves' draw a stem and leaf diagram to show the data.
b) What is the modal class of the data?   **c)** What is the median value of the data?

Displaying Data 1

1 A survey was carried out to find the favourite type of music for a group of people. The results are displayed in this pictogram.

Country
Pop
Classical
Jazz
Rock

Where ⬭ represents four people.

Draw a bar graph to show this information.

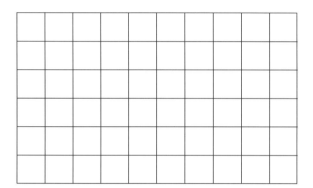

2 36 primary school children were asked to name their favourite pet.

$\frac{1}{3}$ said DOG

$\frac{1}{4}$ said CAT

$\frac{2}{9}$ said FISH

$\frac{1}{12}$ said RABBIT

... and the remainder said BIRD.

Draw a bar graph to show this information.

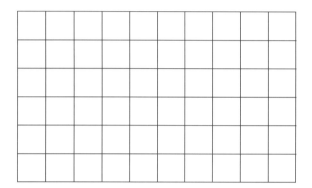

3 This bar chart shows the different types of milk sold in a supermarket and the actual number of pints sold on one day.

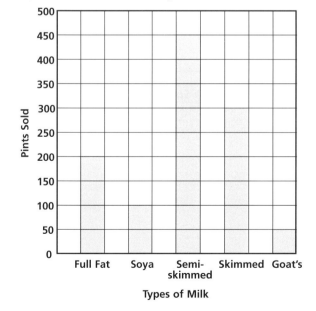

a) What was the second most popular type of milk sold?

....

b) How many pints of milk are sold in total?

....

c) In the space below draw a pictogram to show this data:

4 **Draw a)** a pictogram and **b)** a bar graph to show the data recorded in question **2** on page 104.

5 **Draw a)** a pictogram and **b)** a bar graph to show the data recorded in question **3** on page 104.

Displaying Data 2 & 3

1 A group of factory workers were asked how long it took them to get to work. This table shows the results. Construct a frequency diagram to show this information.

Time, t (minutes)	Number of workers i.e. Frequency
$0 \leqslant t < 10$	21
$10 \leqslant t < 20$	14
$20 \leqslant t < 30$	28
$30 \leqslant t < 40$	8
$40 \leqslant t < 50$	4

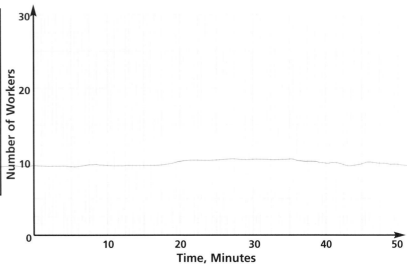

2 In a survey 50 women were asked how much they spend on cosmetics in one week. The results are shown in the table below.

Money Spent, M (£)	Frequency
$0 \leqslant M < 2$	1
$2 \leqslant M < 4$	6
$4 \leqslant M < 6$	28
$6 \leqslant M < 8$	11
$8 \leqslant M < 10$	4

a) Draw a frequency polygon to show the information.

b) 50 men were also asked how much they spend on cosmetics in one week. The results are shown below.

On the same axes draw a frequency polygon to show the money spent by men.

Money Spent, M (£)	Frequency
$0 \leqslant M < 2$	13
$2 \leqslant M < 4$	28
$4 \leqslant M < 6$	5
$6 \leqslant M < 8$	3
$8 \leqslant M < 10$	1

c) How do the two distributions compare?

....

....

....

....

....

....

....

....

....

3 The table shows the height of a girl from birth to age 5. Her height was recorded every year on her birthday.

Height (cm)	42	51	66	75	84	89
Time (years)	0	1	2	3	4	5

a) Draw a time series to show the trend in her height.

b) Between which time period was there the greatest increase in height?

....

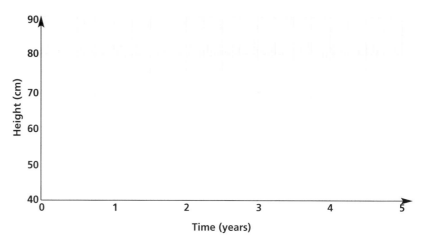

4 The graph below shows the sales of ice cream at a shop at a seaside resort. What trend, if any, does the graph show?

Seasons to winter 2003

....
....
....
....
....
....
....
....
....

5 The table below shows the sales of shoes at a store over a six month period. Use a three point moving average to determine if there is a trend in sales.

Apr	May	Jun	Jul	Aug	Sept
103	78	83	73	57	122

....
....
....
....
....
....

6 Below are the highest recorded temperatures, in °C, on one particular day for forty places around the world.

17, 28, 33,19, 21, 28, 31, 24, 21, 20, 19, 28, 24, 19, 20, 24, 29, 32, 16, 26,
33, 24, 23, 16, 16, 20, 28, 17, 24, 23, 26, 31, 33, 18, 31, 26, 28, 19, 19, 21

a) Group together the data in a frequency table.

b) Construct a frequency diagram to show the data.

c) On separate axes construct a frequency polygon to show the data.

Scatter Diagrams 1 & 2

1 This scatter diagram shows the average journey time and distance travelled for ten pupils travelling from home to school.

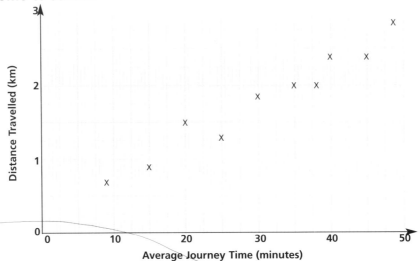

a) What does the scatter diagram tell you about the relationship between the journey time and the distance travelled?

....

b) Draw a line of best fit.

c) Use your graph to estimate ...

i) the time taken by John, who travelled a distance of 2.5 km.   ....

ii) the distance travelled by Donna, who takes 27 minutes. ....

2 The table below shows the heights and weights of 10 boys.

Height (cm)	133	162	130	163	153
Weight (kg)	70	84	64	87	87

Height (cm)	150	124	141	150	138
Weight (kg)	77	66	79	82	69

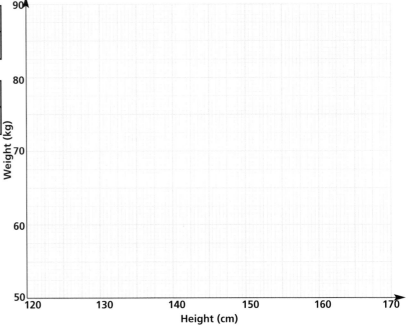

a) Use the information given to plot a scatter diagram, including line of best fit.

b) What type of correlation is there between height and weight?

....

c) i) Tony weighs 72kg. Use your graph to estimate his height. ....

ii) Rob is 1.57m tall. Use your graph to estimate his weight..................................

3 Mrs Thrift goes shopping at her local supermarket on 12 separate occasions. Each time she pays for her items with a £10 note. The table below shows the number of items bought and change received.

| Change received (£) | 2.50 | 5.60 | 5.70 | 7.80 | 0.90 | 3.10 | 5.20 | 4.20 | 1.50 | 7.90 | 6.80 | 2.70 |
| Number of items bought | 10 | 8 | 6 | 4 | 14 | 12 | 7 | 8 | 14 | 2 | 7 | 11 |

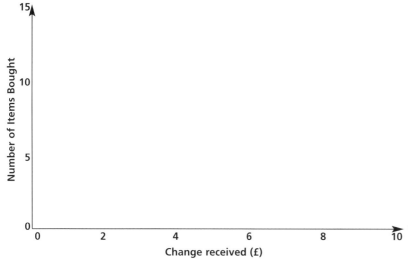

a) Use the information given to plot a scatter diagram including line of best fit.

b) What type of correlation does the scatter diagram show? ....

c) i) Mrs Thrift buys 9 items. Estimate how much change she receives. ....

ii) Mrs Thrift receives £6.20 in change. Estimate how many items she bought.

....

4 The table below shows the number of tracks and total playing time for 12 music CDs.

| Number of tracks | 14 | 20 | 8 | 11 | 18 | 14 |
| Total time (mins) | 62 | 69 | 58 | 61 | 66 | 67 |

| Number of tracks | 10 | 5 | 16 | 7 | 18 | 6 |
| Total time (mins) | 56 | 46 | 66 | 56 | 69 | 52 |

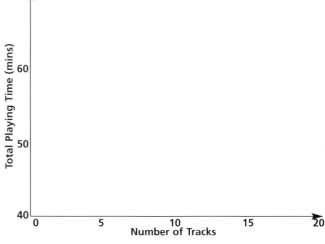

a) Use the information given to plot a scatter diagram including line of best fit.

b) i) If a CD has 13 tracks, estimate its total playing time. ....

ii) If the total playing time of a CD is 54 minutes, estimate the number of tracks it has. ....

5 The table gives information about the number of chapters and the total number of pages in the books on Diane's shelf.

| Number of chapters | 19 | 28 | 11 | 14 | 27 | 23 | 8 | 16 | 21 | 25 | 32 | 19 | 35 | 11 | 16 |
| Total number of pages | 250 | 355 | 110 | 230 | 235 | 350 | 145 | 200 | 235 | 315 | 325 | 120 | 395 | 125 | 305 |

a) Use the information given to plot a scatter diagram including line of best fit.

b) What does the scatter diagram tell you about the relationship between the number of chapters and total number of pages?

c) Use your graph to estimate **i)** the total number of pages if a book has 24 chapters.

ii) the number of chapters if a book has a total of 190 pages.

Pie Charts

1 In one week a travel agent sold 120 separate holidays. The table below shows the holiday destinations.

Holiday Destination	No. of holidays sold
Spain	42
Greece	12
France	34
Cyprus	22
Italy	10

a) Draw and label a pie chart to represent these destinations.

b) What percentage of the holidays sold were for Spain?

....

c) What fraction of the holidays sold were for Greece?

....

2 Rose has completed 40 pieces of work in Maths in one school year. Her grades are as follows:

B, C, A, B, C, D, B, C, B, E, D,
A, B, C, D, B, C, B, B, D, A, C,
C, B, A, D, A, C, B, C, B, A, D,
E, C, B, C, C, B, D.

a) Draw a table to show the distribution of grades.

b) Draw and label a pie chart to show the distribution.

c) What fraction of Rose's grades were A's or B's?

....

3 **60 men and 60 women were asked, 'What is your favourite colour of car?'**

The two pie charts show the results.

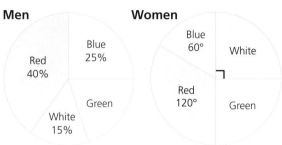

a) i) How many men chose red as their favourite colour?

....

ii) How many women chose red as their favourite colour?

....

b) Did more men or women choose green as their

favourite colour? Show your working.

....

....

4 **This pie chart shows the daily newspaper bought by 72 people.**

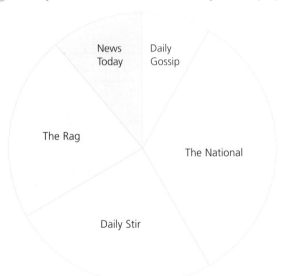

a) i) Which is the most popular newspaper?

....

ii) Which is the least popular newspaper?

....

b) i) How many people buy the Daily Stir?

....

ii) How many more people buy The National

compared to News Today?

....

....

5 **This table shows the number of telephone calls Mrs Chattergee makes in one week.**

Day of the week	Monday	Tuesday	Wednesday	Thursday	Friday	Saturday	Sunday
Number of phone calls	3	2	8	4	7	10	6

a) Draw and label a pie chart to represent the information.
b) What percentage of the telephone calls were made at the weekend?

6 **The ages of the people living in Addick Close are as follows:**
2, 24, 14, 8, 70, 15, 19, 31, 85, 4, 3, 12, 30, 27, 45, 50, 66, 68, 2, 11,
74, 31, 63, 28, 41, 47, 51, 14, 18, 83, 69, 7, 52, 35, 33, 20, 14, 6, 7, 5.
a) Construct a frequency table to show the distribution of ages using class intervals of 1-20, 21-40, etc.
b) Draw and label a pie chart to show the distribution. c) What percentage of the residents are aged 41 or over?

7 **A group of men and women were asked how they travel to work.**
The two pie charts show their responses. They are not drawn accurately.
a) If 10 men travel to work by car, how many men
 travel to work by i) bus? ii) train?
b) If 20 women travel to work by train, how many women
 travel to work by i) car? ii) bus?

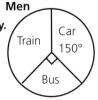

Mean, Median, Mode & Range 1

1 **Laura has the following coins in her pocket:**

Calculate the mean, median, mode and range for the coins.

Mean: ....

....

Median: .... Mode: .... Range: ....

2 **The table below shows the results of an Internet search for the price of a particular camera.**

Calculate the mean, mode, median and range of the camera prices.

Camera	Cost (£)	Camera	Cost (£)	Camera	Cost (£)	Camera	Cost (£)	Camera	Cost (£)
1	227	5	169	9	225	13	211	17	155
2	246	6	204	10	210	14	248	18	153
3	248	7	220	11	239	15	166	19	196
4	248	8	165	12	227	16	170	20	173

Mean: ....

....

Median: .... Mode: .... Range: ....

3 **The weights of the 11 players in the Year 11 hockey team were measured.**

Their weights in Kilograms were: 51, 60, 62, 47, 53, 48, 52, 51, 65, 61, 66

a) Calculate the mean, median, mode and range for their weights.

Mean: ....

....

Median: .... Mode: .... Range: ....

b) The weights of the 11 players in the Year 10 hockey team were also measured.

Their mean was 52.5kg and the range was 11kg. Compare the weights of the two teams.

....

....

4 **9A contains 14 girls. Their heights in metres are:**
1.46, 1.62, 1.57, 1.6, 1.39, 1.71, 1.53, 1.62, 1.58, 1.40, 1.46, 1.63, 1.62, 1.65
a) Calculate the mean, median, mode and range for the heights.
b) The boys in 9A have a mean height of 1.68m and a range of 0.32m. Compare the heights of the girls and the boys.

Mean, Median, Mode & Range 2 & 3

1 Janet carries out a survey on the number of passengers in cars which pass her house.

Here are the results:

Number of passengers	Frequency	Frequency x No. of passengers
0	11	
1	12	
2	6	
3	8	
4	3	

a) How many cars were there in her survey?.................................

b) What is the modal number of passengers? ....

c) What is the median number of passengers?.................................

d) What is the range of the number of passengers? ....

....

e) What is the mean number of passengers?

....

....

2 This graph shows the number of chocolate bars bought by pupils at a school tuck shop.

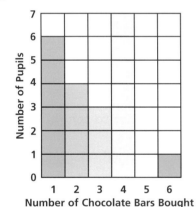

Number of Chocolate Bars Bought

a) How many pupils bought chocolate bars? ....

b) What is the modal number of bars bought? ....

c) What is the median number of bars bought? ....

d) What is the mean number of bars bought?

....

....

....

3 The table gives the recorded midday temperature every day for June.

Temperature, T (°C)	Frequency	Mid-temp values (°C)	Frequency x Mid-temp values
5 ⩽ T < 10	3	7.5	3 x 7.5 = 22.5
10 ⩽ T < 15	14	12.5	
15 ⩽ T < 20	11		
20 ⩽ T < 25	2		

a) Complete the table and work out an estimate of the mean recorded temperature.

b) In which class interval does the median lie?...

c) What is the modal class? ....

4 Phil carries out a survey about the amount of pocket money his classmates each receive every week. The results are shown below.

Amount of pocket money, M (£)	Frequency		
$0 \leqslant M < 2$	3		
$2 \leqslant M < 4$	15		
$4 \leqslant M < 6$	8		
$6 \leqslant M < 8$	5		
$8 \leqslant M < 10$	1		

a) Calculate an estimate of the mean amount of pocket money the pupils receive.

....

....

b) In which class interval does the median lie? ....

c) What is the modal class? ....

5 The graph shows the amount of time Narisha had to wait for a bus to take her to school each morning for half a term.

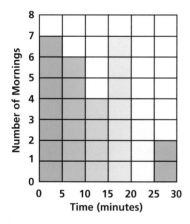

a) How many mornings are shown altogether? ....

....

b) Calculate an estimate of the mean amount of time Narisha had to wait for a bus.

....

....

....

....

6 The table below shows the number of goals scored by the Year 11 football team in one season.

Number of goals scored	0	1	2	3	4	5	6
Number of games	7	5	11	3	3	0	1

a) What is **i)** the mean **ii)** the range **iii)** the mode and **iv)** the median of the number of goals scored?
b) The Year 10 football team averaged 2.2 goals per game in five less games. Which year scored the most goals and by how many?

7 During a PE lesson the boys have a 100m race. Their times were recorded and the results are shown below:

Time taken, t (seconds)	$12 < t \leqslant 14$	$14 < t \leqslant 16$	$16 < t \leqslant 18$	$18 < t \leqslant 20$	$20 < t \leqslant 22$	$22 < t \leqslant 24$
Number of boys	2	9	13	5	3	1

a) Calculate an estimate of the mean time
b) In which class interval does the median lie?
c) Which is the modal class?

Cumulative Frequency 1 & 2

1 The cumulative frequency graph shows the time taken for 30 students to complete their maths homework.

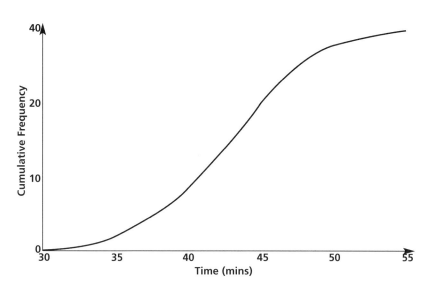

a) How many students took less than 40 minutes?

....

b) How many students took more than 40 minutes?

....

c) What was the median time taken?

....

2 Thirty five people took part in a skateboarding competition. The points they scored are shown in the table below.

Points scored (P)	Frequency	Cumulative Frequency
$0 < P \leqslant 5$	2	
$5 < P \leqslant 10$	4	
$10 < P \leqslant 15$	5	
$15 < P \leqslant 20$	7	
$20 < P \leqslant 25$	12	
$25 < P \leqslant 30$	5	

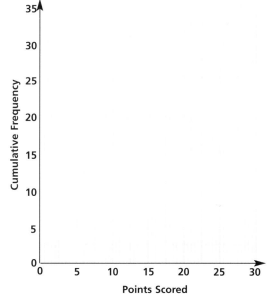

a) Complete the cumulative frequency column in the table.

b) Draw the cumulative frequency graph.

c) Use the graph to estimate:

i) The median points scored ....

ii) The inter-quartile range ....

d) How many competitors scored between 15 and 25 points?

....

e) How many competitors scored more than 25 points?

....

3 The cumulative frequency graph shows

the Resistant Materials test results for 80 boys.

The test was out of 50.

a) What was the median result?

....

b) What was the inter-quartile range?

....

c) How many boys scored more than 45?

....

d) 80 girls also completed the test. The table

below shows their results. Complete the table.

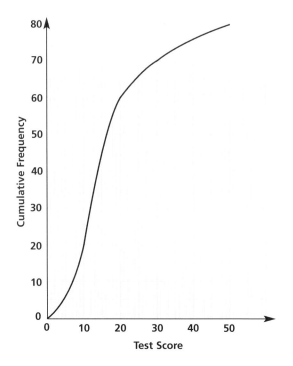

Test Result (T)	Frequency	Cumulative Frequency
0 < T ≤ 10	4	
10 < T ≤ 20	13	
20 < T ≤ 30	19	
30 < T ≤ 40	34	
40 < T ≤ 50	10	

e) Draw the cumulative frequency curve for the

girls on the same graph.

f) What was the median result for the girls?

....

g) How do the two results compare? Use two readings to support your comparisons.

....

....

....

4 The heights in cm of 300 pupils were recorded as shown.
  a) Complete the table, adding a cumulative frequency column.
  b) Draw the cumulative frequency curve.
  c) Use your graph to find: i) Median height ii) Inter-quartile range
  iii) Number of pupils taller than 155cm.

Height (h)	Frequency
130 < h ≤ 140	10
140 < h ≤ 150	39
150 < h ≤ 160	95
160 < h ≤ 170	125
170 < h ≤ 180	31

5 In one week a doctor weighed 80 men.
  The table shows the results:

Weight	50 < W ≤ 60	60 < W ≤ 70	70 < W ≤ 80	80 < W ≤ 90	90 < W ≤ 100
Frequency	11	27	29	8	5

a) Draw a cumulative frequency curve.
b) Use your graph to estimate:
i) Median weight ii) Inter-quartile range
c) What percentage of men weighed more than 70kg?

Notes

Notes

The Essentials of

AQA maths

Linear Specification A(3301) **Intermediate Tier**

H. Rees and P. Wharton

STUDENT WORKBOOK ANSWERS

Lonsdale REVISION GUIDES

Answers to *The Essentials of AQA Maths Linear Specification A (3301) Intermediate Tier.*

NUMBER

Place Value – page 4
Q1: (a) Five thousand three hundred and seven (b) Seventy three thousand, one hundred and fifty eight (c) 560 (d) 1 2 4 7 one thousand, two hundred and forty seven (e) 14052

Q2: (a) 490 000 (b) 494 000 (c) 493 500 (d) 493 510

Q3: (a) 110 000 (b) 109 000 (c) 109 100 (d) 109 110

Q4: 31 500 - 32 500

Q5: (a) 10 (b) 30

Q6:

	10000	1000	100	10	1
(a)					1
(b)				8	7
(c)			4	0	9
(d)		6	2	2	3
(e)	5	0	0	0	5

Q7: (a) 31000 (b) 30700 (c) 30720

Q8: (a) 500,000 (b) Five hundred thousand

Numbers 1 & 2 – page 5
Q1: (a) 4,6,20 (b) 4,20 (c) 3,15 (d) 3,6,7,21 (e) 7,21, (f) 4,6,20 (g) 3,7,11 (h) 4

Q2: (a) 5,11,19,31,47,81 (b) 8,24,36 (c) 5,11,22 (d) 24,36,81 (e) 8,24,36 (f) 5,11,19,31,47 (g) 36,81

Q3: (a) $^1/_8$,0.125 (b) $^1/_{25}$,0.04 (c) 2 (d) $1^1/_3$,1.$\dot{3}$

Q4: 10

Q5: (a) $2^2 \times 3^2$ (b) 2^6 (c) $2 \times 3 \times 5 \times 31$

Q6: (a) p=4; $^1/_2 \times 4 + 1 = 3$ (any multiple of 4 will work) (b) 2+3=5, 2+5=7 (any prime number added to 2 will be odd)

Numbers 1 & 2 (cont) – page 6
Q7: (a) 4 (b) 180

Q8: (a) 4 (b) 1600

Q9: 8 packs of bread, 15 packs of beefburgers

Q10: (a) 12 and 20 (b) 15, 30, 45

Q11: (a)(i)25,50,100 (ii)25,50,100 (iii) 14,70,84 (iv) 29,41,61 (b) 50

Q12: (a)$^1/_{100}$,0.01 (b) 100 (c) 100 (d) $1^1/_{99}$,1.$\dot{0}\dot{1}$

Q13: (a) $2 \times 3 \times 5$ (b) $2^2 \times 5^2$ (c) 2^{11}

Q14: (a) 3,90 (b) 2,1160 (c) 3,360

Q15: 45

Number 3 – page 7
Q1: (a) 503 (b) 28009 (c) 139 (d) 41449 (e) 15228 (f) 428155 (g) 38 (h) 257 (i) 4100 (j) 573 000 (k) 42.3 (l) 0.008

Q2: (a) 234 (b) 8

Q3: (a) 10^5 (100 000) (b) 10^5 (100 000)

Q4: (a) 4000 (b) 5165 (c) 3582 (d) 4293 (e) 31691 (f) 154875 (g) 367 (h) 2076 (i) 9400 (j) 40300 (k) 11000 (l) 339420000 (m) 40.8 (n) 0.55 (o) 0.06 (p) 3.3942

Q5: (a) 230 (b) 10^2

Integers 1 & 2 – page 8
Q1: (a) -11, -7, -3, -1, 2, 5, 12, 14
(b) 467,165,70,8,-62, -162, -230, -320

Q2: (a) (i) 8 (ii) -4 (iii) -5 (iv) 5 (v) 8 (vi) -8 (vii) -2 (viii) -4 (ix) -1 (x) 6 (xi) -4 (xii) -5
(b) (i) 12 (ii) -8 (iii) -16 (iv) -6 (v) -28 (vi) 2
(c) (i) -15 (ii) 16 (iii) 5 (iv) -4 (v) 1
(d) lots of possible answers

Q3: (a) -,= (b) +,-,= (c) x,÷,= (d) ÷,+,= (e) -,+,÷,= (f) +,-,-,=

Integers 1 & 2 (cont) – page 9
Q4: (a) (i) 17°C (ii) 31°C (iii) 4°C (b) 11°C

Q5: (a) £147.30 (b) £260.00

Q6: -36,-11,-3,-1,0,3,4,19,74,100

Q7: (a) -2 (b) -8 (c) 0 (d) 300 (e) -90 (f) 20 (g) -25

Q8: (a) 15 (b) -11 (c) 19 (d) -16 (e) -11

Order of Operations – page 10
Q1: (a) 17 (b) 67 (c) 44 (d) 11 (e) 11 (f) 1 (g) 208 (h) 20 (i) -16

Q2: (a) (13-3)x4+3 (b) 13-3x(4+3) (c) (13-3)x(4+3) (d) 13-(3x4)+3 (e) (3x1.4+4)x2.5 (f) 7+3.2x(6-4.4)

Q3: (a) 43 (b) -1 (c) -35 (d) 11 (e) 12 (f) 10 (g) -1.5 (h) -35

Q4: (a) 3^2-(4x5)+10 (b) $(3^2$-4)x5+10 (c) 3^2-(4x(5+10))

Q5: (a) $(4^2$-3)x4+2 (b) 4^2-3x(4+2) (c) 4^2-(3x4)+2

Rounding Numbers 1 & 2 – page 11
Q1: (a) (i) 7.3 (ii) 7.32 (b) (i) 16.8 (ii) 16.78 (c) (i) 0.02 (ii) 0.018 (d) (i) 0.1 (ii) 0.105 (e) (i) 7.1 (ii) 7.07

Q2: Lowest 68.35kg, Highest 68.45kg

Q3: Lowest 1.645m, Highest 1.655m

Q4: (a) (i) 400 (ii) 430 (b) (i) 9000 (ii) 9200 (c) (i) 10000 (ii) 10050 (d) (i) 0.02 (ii) 0.024 (e) (i) 0.00017 (ii) 0.000174 (f) (i) 0.01 (ii) 0.0104

Q5: 5750 – 5650 = 100

Q6: (a) 135.7 (b) 135.67 (c) 100 (d) 140 (e) 136

Q7: (a) 0.1 (b) 0.06 (c) 0.060 (d) 0.06 (e) 0.060 (f) 0.0604

Q8: 47.65 – 47.55 = 0.1kg

Q9: (a) 14.303036 (b) 14.3 (3 s.f.) or 14.30 (2 d.p.)

Estimating & Checking – page 12
Q1: (a) (i) 5618 (ii) 100 x 50 = 5000 (b) (i) 57.76 (ii) 4 x 15 = 60
(c) (i) 8.7 (ii) $\frac{30 \times 6}{20}$ = 9 (d) (i) 273.15546 (ii) 400000 x 0.0007 = 280

Q2: (a) (i) 7 + 9+13 = £29 (ii) £29.73 John accurate (b) (i) 9+17+5 = £31 (ii) £30.20 Donna inaccurate

Q3: (a) 10+30+40 = 80 inaccurate 85 (b) 9+12+7+14 = 42 accurate 41.73 (c) (3x6)+90-10 = 98 inaccurate 100.75 (d) (5x6)+(1x10)=40 inaccurate 37.21

Q4: (a) 6.692307692 (b) $\frac{9}{3-2}$ = 9 (c) 6.7 (2 s.f.)

Powers 1 & 2 – page 13
Q1: (a) 8 (b) 9 (c) 64 (d) 1000 (e) 81 (f) 1

Q2: (a) 25,64,196 (b) 27,64,125 (c) 64 (d) 64 (e) 80

Q3: (a) 9 (b) -27 (c) 25 (d) -125 (e) 1 (f) -1

Q4: (a) 32 (b) 27 (c) 4096 (d) 2 (e) 100 (f) 1296 (g) 16 (h) 4096 (i) 1 000 000 (j) 1024 (k) 1 (l) 2 (m) 1 (n) 2

Q5: (a) 3^4 (b) 2^7 (c) 4^5

Q6: (a) 256 (b) 4 (c) 1024 (d) 1600 (e) 28 (f) 27 (g) 2 (h) 5 (i) 2

Q7: (a) 3^7 (b) 2^4

Q8: (a) 125 (b) (i) 6.172839506 (ii) 6

Roots – page 14
Q1: (a) ±5 (b) ±6 (c) ±12 (d) ±14 (e) 4 (f) 1 (g) 2 (h) 10 (i) 3

Q2: (a) 4,36,64 (b) 8,27,64 (c) 64 (d) 10 (e) 36

Q3: (a) 4 (b) 4 (c) 20 (d) 30 (e) 32 (f) 9 (g) 64

Q4: (a) 4^2 (b) 3^2 (c) 2^4 (d) 6^2

Q5: (a) 3√3 (b) 5√2 (c) 3√6 (d) 3√10

Q6: (a) 13 (b) 5 (c)9 (d) 5 (e) 15

Q7: (a) 27 (b) 4 (c)4 (d) 8 (e) 16

Q8: (a) 10√2 (b) 4√5 (c) 3√7

Standard Form – page 15
Q1: (a) 230 (b) 2300 (c) 42 100 (d) 63.2 (e) 7467 (f) 0.03 (g) 0.0023 (h) 0.000421 (i) 0.06324

Q2: (a) 6×10^2 (b) 4.73×10^2 (c) 4.2×10^4 (d) 4.13256×10^5 (e) 4.963×10^2 (f) 3.2×10^{-2} (g) 4.7×10^{-1} (h) 6.31×10^{-4} (i) 1×10^{-1}

Q3: (a) 9.8×10^6 grams (b) 4×10^{-5} grams (c) 2.45×10^{11} grains

Q4: (a) 5.633×10^3 (b) 6.182×10^5 (c) 8.424×10^{11} (d) 5×10^{-2}

Q5: 220

Q6: (a) 1.08×10^{12} (b) 9.4608×10^{15} (c) 8 mins (d) 4.068144×10^{16}

Q7: (a) (i) 1.05×10^7 (ii) 9.6×10^{-3} (b) 0.0096 (c) 3.2×10^7

Fractions 1 & 2 – page 16
Q1: (a) $^{16}/_{40}$, $^{40}/_{100}$, $^{10}/_{25}$ (b) $^{35}/_{50}$, $^{28}/_{40}$, $^{84}/_{120}$

Q2: (a) $^9/_{10}$ (b) 7 (c) $^4/_5$ (d) $^{27}/_{46}$

Q3: (a) $^1/_2$, $^3/_5$, $^2/_3$, $^{11}/_{15}$, $^5/_6$ (b) $^9/_{10}$, $^5/_8$, $^3/_5$, $^1/_4$, $^9/_{40}$

Q4: $^{11}/_{15}$, $^{23}/_{30}$, $^{12}/_{15}$, (any two)

Q5: (a) (i) $2^1/_5$ (ii) $2^1/_6$ (iii) $4^4/_5$ (iv) 10 $^2/_3$
(b) (i) $^{10}/_3$ (ii) $^{21}/_4$ (iii) $^{58}/_5$ (iv) $^{40}1/_{20}$

Q6: (a) $^4/_6$, $^6/_9$, $^8/_{12}$, $^{10}/_{15}$ etc. (b) $^4/_7$, $^8/_{14}$, $^{12}/_{21}$, $^{16}/_{28}$ etc. (c) $^9/_{11}$, $^{18}/_{22}$, $^{27}/_{33}$, $^{36}/_{44}$ etc.

Q7: $^1/_2$, $^3/_4$, $^4/_5$, $^9/_{10}$, $^{19}/_{20}$

Q8: $^{41}/_{50}$, $^{42}/_{50}$, $^{43}/_{50}$, $^{44}/_{50}$, $^{33}/_{40}$, $^{34}/_{40}$, $^{35}/_{40}$ (any three)

Fractions 3 – page 17
Q1: (a) $1^5/_{12}$ (b) $1^7/_{72}$ (c) $7^2/_5$ (d) $11^{23}/_{24}$ (e) $^2/_5$ (f) $^1/_5$ (g) $3^{17}/_{40}$ (h) $4^{17}/_{30}$

Q2: $^1/_3$

Q3: (a) $^1/_{10}$ (b) $^9/_{35}$ (c) $^5/_6$ (d) $1^1/_2$ (e) $5^3/_5$ (f) $5^4/_7$ (g) $1^{11}/_{16}$ (h) $^2/_3$ (i) $3^5/_9$ (j) $^7/_{30}$

Q4: $^1/_5$ + $^1/_3$ = $^3/_{15}$ + $^5/_{15}$ = $^8/_{15}$

Q5: (a) $1^{13}/_{40}$ (b) $6^{17}/_{24}$ (c) $^7/_{30}$ (d) $^4/_5$ (e) $^7/_{25}$ (f) $1^1/_{12}$ (g) $5^5/_{12}$ (h) $1^1/_5$ (i) $22^1/_2$ (j) $^7/_{18}$

Calculations Involving Fractions – page 18
Q1: £30

Q2: (a) £5200 (b) £162.50 (c) $^{11}/_{20}$

Q3: (a) £78.00 (b) £145.80

Q4: $^5/_6$

Q5: (a) (i) 300 (ii) £225 (b) £400
Q6: $^1/_{90}$

Decimals 1 & 2 – page 19
Q1: (a) 102.5 (b) 13.471 (c) 8.407 (d) 90.031 (e) 423.008
Q2: (a) 0.6 (b) 0.4 (c) 0.09 (d) 0.7
Q3: 0.306, 0.36, 0.63, 3.6, 6.3
Q4: (a) 21.39 (b) 108.037 (c) 32.46 (d) 717.21
Q5: (a) 0.375 (b) 0.2 (c) 0.03 (d) 0.4 (e) 0.26
Q6: 143.2, 14.32, 14.23, 13.42, 1.432, 1.342
Q7: £4.91
Q8: 28.16kg

Multiplication of Decimals – page 20
Q1: (a) 47 (b) 132.46 (c) 0.146 (d) 136300 (e) 98.28 (f) 10.856
(g) 286.488 (h) 7.35228
Q2: (a) 97.2 (b) 9.72 (c) 9.72 (d) 0.972 (e) 0.0972
Q3: (a) 195.657 (b) 19565700 (c) 195657 (d) 19565.7
Q4: (a) 156.7 (b) 1.01 (c) 3467.1 (d) 25.52 (e) 8.406 (f) 358.19
(g) 3.0659 (h) 0.0282
Q5: (a) 16544 (b) 165.44 (c) 1.6544 (d) 0.16544
Q6: £109.97
Q7: £47
Q8: £39.92
Q9: 100cm

Division of Decimals – page 21
Q1: (a) 1.63 (b) 0.0347 (c) 146.324 (d) 0.0012467 (e) 4.7 (f) 34.2
(g) 4964 (h) 458
Q2: (a) 13 (b) 0.13 (c) 130 (d) 0.013 (e) 13
Q3: (a) 34 (b) 270 (c) 3.4 (d) 2.7 (e) 0.27
Q4: (a) 0.7162 (b) 0.000034 (c) 0.4731 (d) 78.6 (e) 17.6 (f) 67.6
Q5: (a) 17.3 (b) 17300 (c) 3.4 (d) 58.82
Q6: 60 tickets
Q7: 14 days

Percentages 1 – page 22
Q1: (a) 12p (b) 1.95km
Q2: (a) 20% (b) 5%
Q3: (a) (i) 0.54m (ii) 45% (b) (i) 24kg (ii) 60%
Q4: 12.5%
Q5: (a) £1.17 (b) 5.12kg (c) £0.52
Q6: (a) 66.67% (2 d.p.) (b) 0.1% (c) 27.5%
Q7: 30%
Q8: 17.5%

Percentages 2 – page 23
Q1: £11550
Q2: Mr Dixon £9.60, Mrs Asaf £8.50. Mr Dixon's is greater.
Q3: (a) £145200 (b) 1.21
Q4: (a) £5120 (b) 0.512
Q5: No, 10% off discounted price is £72
Q6: (a) 113.4kg (b) 0.945

Percentages 3 – page 24
Q1: 40 mins
Q2: £144
Q3: £4500
Q4: 220km
Q5: £60000
Q6: £124
Q7: £3200
Q8: 300 stamps

Converting Between Systems – page 25
Q1: (a) 0.3, 30% (b) $^9/_{20}$, 45% (c) $^3/_8$, 0.375 (d) 0.6, 66.6% (e) $^1/_8$, 12.5%
(f) $^{21}/_{25}$, 0.84 (g) 1.8, 180% (h) $4^3/_5$, 460% (i) $2^1/_4$, 2.25
Q2: $^3/_5$ = 60% MORE THAN 55%
Q3: 0.25, $^1/_3$, 35%, $^2/_5$, 0.42, 44%
Q4: 70% of £50 = £35, $^4/_{10}$ of £90 = £36, the largest
Q5: (a) (i) 0.8 (ii) 0.9 (iii) 0.85 (iv) 0.85 (b) $^{19}/_{20}$, 0.92, 90%, 85%, $^4/_5$, 0.75
Q6: (a) (i) 62% (ii) 60% (iii) 56% (iv) 58%
(b) 0.56, $^{29}/_{50}$, $^3/_5$, 61%, 0.62, 65%

Everyday Maths 1 & 2 – page 26
Q1: (a) £446.50 (b) £440
Q2: (a) Simple £22 000, Compound £21 966.08, Simple earns most
(b) Simple £25 000, Compound £25 283.45, Compound earns most
Q3: (a) £2400 (b) 6.5% increase

Everyday Maths 1 & 2 (cont) – page 27
Q4: (i) 2190 (ii) £175.20 (iii) £184.80 (iv) £9.24 (v) £194.04
Q5: (a) 11:02, 53 mins (b) 10:47 (c) 60%
Q6: £211.50
Q7: Simple £23600 Compound £23484.83   Simple earns more
Q8: 30%
Q9: 883 units
Q10: 2.5 mins

Ratio and Proportion 1 & 2 – page 28
Q1: (a) 70:35   (b) 1:0.5
      = 2:1
Q2: (a) 90:135   (b) 1:1.5
      2:3
Q3: (a) 3:4 (b) 1: $^4/_3$ or 1:1.3
Q4: (a) 9:5 (b) 18:11 (c) Larger Tin (Large 1p = 12.5g, Small 1p = 11.4g)
Q5: (a) 32p: 48p (b) 140cm: 210cm: 280cm
Q6: Susan – 60 years old, Janet – 36 years old, Polly – 24 years old

Ratio and Proportion 1 & 2 (cont) – page 29
Q7: 14:10 or 7:5
Q8: 2778kg
Q9: (a) 225g flour, 325g oatmeal, 200g margarine (b) 55 biscuits
Q10: (a) 4.8:1.8 or 8:3
(b) 1: $^3/_8$ or 1:0.375
Q11: (a) Large 1: 4.54g Small 1: 5.28g Small best value
Q12: Lucy £3000, Paul £3500, John £4000, Sarah £4500
Q13: £65
Q14: 160°
Q15: £1.68

ALGEBRA
The Basics of Algebra 1 & 2 – Page 30
Q1: (a) 3a (b) 11χ (c) 5a (d) 11p (e) $5a^2$ (f) $11w^2$ (g) 6a+8b (h) -2c + 2d
(i) 6ab (j) 24pqr (k) -2ab + 9cd (l) $-χ^2$-5χ+9 (m) 17– $6p^2$+7p
Q2: χ can take any value in an identity.
Q3: (a) a^5 (b) $12b^4$ (c) $24p^{10}$ (d) r^4 (e) 4c (f) 2ab (g) $5a^2b$ (h) $8a^6b^3$
Q4: (a) 2ab (b) $16a^2$ (c) 8a (d) $9b^2c$
Q5: (a) 18χ (b) 6χ (c) $72χ^2$ (d) 2 (e) $-7p^2$ (f) 21ab +3bc
(g) 9ab + $7a^2b$ + $8ab^2$
Q6: (a) a^9 (b) $30χ^8$ (c) $χ^2$ (d) 2abc (e) $64a^6$ (f) $81a^8b^4$

Substitution – page 31
Q1: (a) 19 (b) 14 (c) 20 (d) 20 (e) -15 (f) -40 (g) 24 (h) -2 (i) 98 (j) $-^2/_5$
Q2: (a) $^5/_6$ (b) -1 (c) 2 (d) 0
Q3: (a) -16 (b) 32 (c) 64 (d) 24 (e) 60
Q4: (a) 24 (b) -32 (c) -1 (d) 7 (e) $4^1/_4$ (f) 1 (g) 4 (h) -2 (i) 23
(j) 1 (k) $^7/_{12}$ (l) 33
Q5: (a) 77 (b) -35 (c) 17 (d) 15 (e) -24
Q6: (a) $1.5×10^8$ (b) $3.5×10^4$ (c) $4.3×10^3$

Brackets and Factorisation – page 32
Q1: (a) 8χ+4 (b) 9r-21 (c) $6m^2$+4m (d) 8p–$2p^2$ (e) $6r^3$–18r
(f) $40χ^3$–$30χ^2$ (g) 10y+23 (h) 17χ–$15χ^2$ (i) 46χ–$33χ^3$ (j) $19χ^2$–2 (k) $42χ^2$–1
(l) $10χ^2$–27χ+4 (m) $χ^2$+7χ+10 (n) $2χ^2$+11χ+12 (o) $6χ^2$–χ-2
(p) $4χ^2$–21χ+5 (q) $χ^2$ – 9 (r) $16χ^2$–8χ+1
Q2: (a) 5(χ+2) (b) 4(χ-2) (c) 2(3χ+ 5y) (d) 3χ(2+χ) (e) 5χ(2χ-1)
(f) 2χy(3χ– 5) (g) 2pqr($2pq^2$+3) (h) (χ+z)(3y+2) (i) (4y+3)(χ+(4y+3))
Q3: (a) 24χ-6y (b) $6χ^2$ + 15χ (c) $8χ^2y$–$4χ^3$ (d) 18χ– $6χ^3$ (e) 13χ– 23
(f) $12χ^2$–14χ+ 20 (g) $40χ^3$+$18χ^2$+ 6χ (h) $12χ^2$+14χ+ 4 (i) $25χ^2$–36
(j) $8χ^3$+$14χ^2$+6χ (k) $8χ^3$–$6χ^5$ –12+$9χ^2$
Q4: (a) 3(3χ-5) (b) 9(χ+1) (c) χ(20χ-1) (d) 4χ(1-$5χ^2$) (e) 4χy(5χy+9)
(f) (3χ-5)(4+y)
Q5: (a) a^2+2ab + b^2 (b) 100

Solving Linear Equations 1 & 2 – page 33
Q1: (a) χ=7 (b) χ=4 (c) χ=3 (d) χ=4 (e) χ=5 (f) χ=$-^1/_2$ (g) χ=18 (h) χ=15
(i) χ=13 (j) χ=3 (k) χ=1.5 (l) χ=-8 (m) χ=$^1/_2$ (n) χ=4.5 (o) χ=3
(p) χ=$^3/_2$ or $1^1/_2$ (q) χ=-4 (r) χ=3
Q2: (a) 3χ+12 =27 (b) $^χ/_4$+11=13
         χ=5   χ=8

Solving Linear Equations 1 & 2 (cont) – page 34
Q3: (a) χ+2χ+2χ+4 (b) 5χ= 64-4
        = 5χ+4   χ=12
                         Helen is 28 years old
Q4: (a) 4χ+2cm   (b) 4χ+2=50
                       χ=12cm
Q5: (a) 6(χ+5) or 6χ + 30 (b) 6(χ+5)=72
                              χ=7cm

2

Q6: (a) $\chi=6$ (b) $\chi=-4$ (c) $\chi=6$ (d) $\chi=-5$ (e) $\chi=-7.5$ (f) $\chi=-4$ (g) $\chi=6$ (h) $\chi=3$
(i) $\chi=2$ (j) $\chi=4$ (k) $\chi=2.5$ (l) $\chi=3.5$ (m) $\chi=1.4$ (n) $\chi=3$ (o) $\chi=-1$ (p) $\chi=2$
Q7: $3(\chi-6)=21$
$\chi=13$
Q8: (a) $3\chi=180°$ (b) $17\chi+20=360$ (c) $^{20}/_3\chi+20=180$
$\chi=60°$ $\chi=20°$ $\chi=24°$

Formulae 1 & 2 – page 35
Q1: (a) $\chi=3-4y$ (b) $\chi=\dfrac{50-7y}{6}$ (c) $\chi=\dfrac{3y+5}{7}$ (d) $\chi=\dfrac{10-6y}{3}$ (e) $\chi=20-3y$
(f) $\chi=\dfrac{8y+3}{4}$ (g) $\chi=\dfrac{18-3y}{8}$ (h) $\chi=\sqrt{3y}{4}$ (i) $\chi=\sqrt{18y}$ (j) $\chi=\dfrac{\sqrt{7y-3}}{5}$
Q2: $v=160cm^3$

Formulae 1 & 2 (cont) – page 36
Q3: $s=44m$
Q4: $100°C$
Q5: (a) $\dfrac{A=C^2}{4\pi}$ (b) $A=130cm^3$
Q6: (a) $p=\dfrac{4q+3}{4}$ (b) $p=\dfrac{-q+4}{6}$ (c) $p=\dfrac{11q-15}{2}$ (d) $p=\sqrt{8q-4}{3}$ (e) $p=\dfrac{\sqrt{8q+6}}{3}$
Q7: $A=1.08m^2$
Q8: $r=5.6cm$
Q9: (a) $s=1.5w+2.5$ (b) £8.50

Factorising Quadratic Expressions – page 37
Q1: (a) $(\chi+4)(\chi+2)$ (b) $(\chi+5)(\chi+2)$ (c) $(\chi+3)(\chi+3)$ (d) $(\chi-4)(\chi-5)$
(e) $(\chi+10)(\chi-1)$ (f) $(\chi+10)(\chi-2)$ (g) $(\chi-4)(\chi+3)$ (h) $(\chi-12)(\chi+2)$
(i) $(\chi-3)(\chi+3)$ (j) $(\chi-8)(\chi+8)$ (k) $(\chi-10)(\chi+10)$ (l) $(\chi-12)(\chi+12)$
Q2: (a) $(\chi+3)(\chi+2)$ (b) $(\chi-3)(\chi-2)$ (c) $(\chi+6)(\chi-1)$ (d) $(\chi-6)(\chi+1)$
(e) $(\chi+6)(\chi-5)$ (f) $(\chi+30)(\chi+1)$ (g) $(\chi-9)(\chi+9)$ (h) $(\chi-13)(\chi+13)$

Solving Quadratic Equations – page 38
Q1: (a) $\chi=-3, \chi=-8$ (b) $\chi=10, \chi=3$ (c) $\chi=-18$ $\chi=2$ (d) $\chi=6, \chi=2$
(e) $\chi=3, \chi=-3$ (f) $\chi=0, \chi=25$
Q2: (a) $\chi^2+4(2\chi)=\chi^2+8\chi$ (b) $\chi=5cm$
Q3: (a) $\chi=-11, \chi=-1$ (b) $\chi=3, \chi=-7$ (c) $\chi=3, \chi=9$ (d) $\chi=10, \chi=4$ (e) $\chi=-8$,
$\chi=5$ (f) $\chi=-11, \chi=2$ (g) $\chi=11, \chi=-2$ (h) $\chi=7, \chi=-6$ (i) $\chi=0, \chi=-9$ (j) $\chi=8, \chi=-2$
(k) $\chi=5, \chi=3$ (l) $\chi=-10, \chi=7$ (m) $\chi=1, \chi=-1$ (n) $\chi=10, \chi=-10$
Q4: (a) $=2(^5/_2\chi)+2(5\chi)+2(^1/_2\chi^2)$ (b) $(\chi+17)(\chi-2)$
$=15\chi+\chi^2$ $\chi=2cm$
(c) $V=(^1/_2\chi)(\chi)(5)$ (d) $v=10cm^3$
$=2.5\chi^2$

Trial & Improvement – page 39
Q1: $\chi=2.6$
Q2: $\chi=3.23$
Q3: (a) $(\chi+3)\chi^2$ (b) $\chi=2.7cm$
$=\chi^3+3\chi^2$
Q4: (a) $\chi=1.8$ (b) $\chi=4.52$
Q5: (a) $(2\chi)(\chi)(\chi+3)=2\chi^3+6\chi$ (b) $\chi=2.2cm$

Number Patterns & Sequences 1 & 2 – page 40
Q1: (a) (i) 31, 63 (ii) -17, -33 (b) (i) 18,34 (ii) -14, -30
Q2: (a) 25, 36, 49, 64 (b) 15, 21, 28, 36
Q3: (a) 23 (b) 58
Q4: (a) -3 (b) 21
Q5: (a) 14th (b) 2nd
Q6: (a) 9th (b) 12th

Number Patterns & Sequences 1 & 2 (cont) – page 41
Q7: $2n+1$
11,13,15,17,19
Q8: $-2n+8$
-2,-4,-6,-8,-10
Q9: (a) $s=4n-3$ (b) 29 (c) 13th diagram
Q10: (a) $c=3n$ (b) 45 (c) 27
Q11: (a)

(b)
```
        1              1
       1 1            1 1
      1 2 1          1 2 1
     1 3 3 1        1 3 3 1
    1 4 6 4 1      1 4 6 4 1
                 1 5 10 10 5 1
```

Q12: (a) 11 (b) 22nd
Q13: (a) $-4n+19$ (b) (i) -21 (ii) -381
Q14: n^2+1

Plotting Points – page 42
Q1: A(3,1) B(3,-1) C(-2,-3) D(-5,5) E(2,0) F(0,-1) G(-3,-1) H(0,4) I(5,-4)
J(-2,0) K(-4,1) L(-3,-4)
Q2: (a) D(0,3) (b) S(-3,-2) (c) H(-2,4)
Q3: D(-3,3)
Q4: S(3,-4)

Graphs of Linear Functions 1 – page 43
Q1: See page 9
Q2: (a) $y=^1/_2\chi+4$ (b) See page 9
(c) $p=6.5$ (d) $q=-8$
Q3: (a) $y=3\chi-4$ (b) See page 9
(c) $y=0.5$ (d) $\chi=2.4$
Q4: (a) See page 9 (b) $\chi=2.5$
Q5: $(-^1/_2,-2)$ See page 9

Graphs of Linear Functions 2 – page 44
Q1: (a) gradient 2, intercept (0,1) (b) gradient 3, intercept (0,-1)
(c) gradient -1, intercept (0,3) (d) gradient -2, intercept (0,1.75)
(e) gradient 1, intercept (0, 4.5) (f) gradient 2, intercept (0,2)
Q2: Line 1: $y=-\chi+3$ Line 2 : $y=3\chi-1$ Line 3: $y=2\chi+1$ Line 4: $y=2\chi+2$
Q3: (a) $y=2\chi+1$ (b) $y=2\chi+1$ (c) $y=2\chi-2$ (d) $y=\chi-2$ (e) $y=-\chi+3$ (f) $y=-\chi-3.5$
Q4: (a) gradient -3, intercept (0,-3) (b) gradient $^3/_4$, intercept (0,2)
(c) gradient $^1/_2$, intercept (0,1.5) (d) gradient $^1/_2$, intercept (0,-1.5)
(e) gradient $^1/_2$, intercept (0,1.5) (f) gradient 2, intercept (0,15)
(g) gradient -4, intercept (0,5)
Q5: (a) $y=\chi+4$ (b) $y=\chi$
Q6: (i),(iii) and (iv)

Graphs of Linear Functions 3 – page 45
Q1: (a) gradient 1, intercept $2y=\chi+2$ (b) gradient 5, intercept $-2y=5\chi-2$
(c) gradient 2, intercept $2y=2\chi+2$ (d) gradient -1, intercept $5y=-\chi+5$
Q2: (a) A (-6,0) B (0,3) (b) $^1/_2$
Q3: (a) (i)1 (ii)-1 (b) (i)$y=\chi+1$ (ii)$y=-\chi+5$
Q4: $y=^1/_2\chi+2$
Q5: $y=-\chi+5$

Three Special Graphs – page 46
Q1: Line 1: $y=-3$ Line 2: $y=\chi$ Line 3: $\chi=-4-6$ Line 4: $y=4$ Line 5: $\chi=1$ Line 6: $y=-\chi$
Q2: (a) $y=0$ (b) $\chi=-1$ (c) $y=-\chi$
Q3: $y=\chi$
Q4: (a) (-3,-3) (b) (-2,6)
Q5: $y=3.6$
Q6: $\chi=-2.8$

Linear Inequalities 1 & 2 – page 47
Q1: (a) $\chi>9$ (b) $\chi<17$ (c) $\chi\leq5$ (d) $\chi\leq2$ (e) $\chi\leq3$ (f) $\chi<3$

9 10 11   15 16 17   3 4 5   0 1 2   1 2 3   1 2 3

Q2: (a) $\chi<-2$ (b) $\chi>-3$ (c) $\chi\geq-1$ (d) $\chi\geq1.5$ (e) $\chi\geq2.5$ (f) $\chi\geq-2.5$
Q3: (a)(b)(c)(d) See page 9

Linear Inequalities 1 & 2 (cont) – page 48
Q4: (a) See page 9 (b) (2,3)
Q5: (a) See page 9 (b) (3,2)
Q6: See page 9
Q7: $y<4$, $\chi\geq-1$, $y\geq\chi$, $\chi+y\geq2$
Q8: (a) $\chi\geq5$ (b) $\chi>2$ (c) $\chi\leq5.5$ (d) $\chi<-2$

5 6 7 8   2 3 4   3 4 5 6   -5 -4 -3 -2

(e) $\chi<4$ (f) $\chi\geq-2$ (g) $\chi<4$

1 2 3 4   -2 -1 0 1   1 2 3 4

Q9: See page 9
Q10: See page 9

Simultaneous Equations 1 – page 49
Q1: (a) $\chi=3$ (d) $\chi=6$
$y=2$ $y=1$
(b) $\chi=7$ (e) $\chi=4$
$y=2$ $y=3$
(c) $\chi=5$ (f) $\chi=3$
$y=-2$ $y=1$
Q2:
$2d+c=84$
$3d+2c=138$
$c=24p$
$d=30p$

Q3: (a) χ=2 (d) χ=-3 (g) χ=-3
    y=4   y=4   y=-2

(b) χ=3 (e) χ=1.5 (h) χ=1.5
    y=6   y=3.5   y=-7.5

(c) χ=4 (f) χ=-3
    y=-2   y=-0.5

Simultaneous Equations 2 – page 50
Q1: (a) See page 9
(b) χ=3
   y=11
Q2: (a) See page 10
(b) χ=3
   y=19
Q3: (a) See page 10
(b) (i) χ=-0.8
    y=1.2

(ii) χ=1
    y=3

(iii) χ=4
    y=0
Q4:(a) See page 10
(b) χ=2 y=9
Q5:(a) See page 10
(b) (i) χ=2 y=4 (ii) χ=5 y=1 (iii) χ=-1 y=-5

Graphs of Quadratic Functions 1 & 2 – page 51
Q1: (a) See page 10
(b) (i) y=0.25 (ii) χ= ±1.8/1.9
Q2: (a)

χ	-3	-2	-1	0	1	2
χ²	9	4	1	0	1	4
+χ	-3	-2	-1	0	1	2
-3	-3	-3	-3	-3	-3	-3
y	3	-1	-3	-3	-1	3

(b) See page 10
(c) (i) y=0.7/0.8 (ii) χ=±1.7
(d) χ=1.3 (e) χ=1.6 (f) χ=0.5
   χ=-2.3   χ=-2.6   χ=-1.5

Graphs of Quadratic Functions 1 & 2 (cont) – page 52
Q3: (a)

χ	-4	-3	-2	1	0	1	2
χ²	16	9	4	1	0	1	4
+2χ	-8	-6	-4	-2	0	2	4
-4	-4	-4	-4	-4	-4	-4	-4
y	4	-1	-4	-5	-4	1	4

(b) See page 10
(c) (i) χ=1.2 (ii) χ²+4χ-6=3 (iii) χ²+4χ-6=-8
   χ=-5.1   χ=1.6   χ=-0.6
         χ=-5.6   χ=-3.4

Q4: (a)

χ	-6	-5	-4	-3	-2	-1	0	1	2-
y	6	-1	-6	-9	-10	-9	-6	-1	6

(b) See page 10
(c) (i) χ=-2.4
   χ=3.2

(ii) χ=-2
   χ=1.6

(iii) χ=-2.8
   χ=3.8

Q5: (a) See page 10
(b) (i) χ=±3.5 (ii) χ=±3.9 (iii) χ=±2.4
Q6: (a) See page 10
(b) (i) χ=1.2 (ii) χ²+2χ-4=2 (iii) χ²+2χ-4=-3
   χ=-3.2   χ=1.6   χ=0.4
         χ=-3.6   χ=-2.4

Graphs of Other Functions – page 53
Q1: (a)

χ	-3	-2	-1	0	1	2
χ³	-27	-8	-1	0	1	8
+10	10	10	10	10	10	10
y	-17	2	9	10	11	18

(b) See page 10
(c) (i) χ=-2.7 (ii) y=13.4
Q2: (a) See page 10
(b) (i) 0.4 (ii) -0.7
Q3: (a)

χ	-3	-2	-1	0	1	2	3
χ³	-27	-8	-1	0	1	8	27
-2	-2	-2	-2	-2	-2	-2	-2
y	-29	-10	-3	-2	-1	6	25

(b) See page 10
(c) χ=1.3
(d) (i) χ=2.8 (ii) y=1.4

Other Graphs 1 – page 54
Q1: (a) 12.30pm (b) Mr Brown 2 hours (c) Mr Smith 60mph
(d) They passed each other (e) Mr Smith 30 mins
Q2: (a) 72km/hr (b) See page 10
Q3: (a) See page 10 (b) 5m/s²
Q4: See page 10
Q5: See page 10

Other Graphs 2 – page 55
Q1: (a)

miles	0	10	20	30	50	100
cost £	30	33	36	39	45	60

(b) See page 11 (c) £46
Q2:   (a) B
      (b) A
      (c) D
      (d) C

Q3: (a)

pounds	10	20	30	40
US Dollars	14	28	42	56

(b) See page 11 (c) £21
Q4: (a) See page 11
(b) Plumber A (c) Plumber B (d) After 4 hours
Q5: See page 11

SHAPE, SPACE & MEASURES
Angles 1 & 2 – page 56
Q1: (a) p = 120° Angles on line add up to 180°
q = 95° Angles on line add up to 180°
(b) p = 144° Angles on line add up to 180° q = 54° vertically opposite
angles (c) p=105° Angles in straight line q = 105° Vertically opposite angles
(d) p = 42° Angles in straight line q = 48° Angles in straight line
(e) p = 38° Vertically opposite angles q = 97° Angles in straight line
(f) p = 112° Vertically opposite angles q = 34° Angles in straight line
Q2: (a) c = 40° Vertically opposite angles d = 40° Alternate angles
(b) c = 37° Verticallly opposite angles d = 37° Corresponding angles
(c) c = 134° Corresponding angles d = 134° Vertically opposite angles

Angles 1 & 2 (cont) – page 57
Q3: (a) m = 62° Alternate angles n = 70° Corresponding angles
(b) m = 85° Allied angles n = 70° Corresponding angles
(c) m = 65° Alternate angles n = 50° Allied angles
Q4: a = 70° Allied angles b = 70° Alternate angles
c = 110° Corresponding angles
Q5: p = 64° Corresponding angles q = 116° Allied angles
r = 64° Alternate angles s = 64° Corresponding angles
Q6: (a) χ = 60° (b) χ = 20° (c) χ = 36° (d) χ = 36°
Q7: (a) p = 110° Alternate angles q = 145° Allied angles
(b) p = 78° Corresponding and alternate angles
q = 48° Straight line angles (c) u = 75° Corresponding angles
v = 75° Alternate angles w = 75° Corresponding angles

Triangles – page 58

Q1: a = e Corresponding b = d Alternate c+d+e = 180° a+b+c = 180°
Q2: Ditto above
Q3: (a) a = 180° – b (b) a = d+f (c) a = 360° – (c+e) (d) 360°

Types of Triangle – page 59

Q1: (a) χ = 70° Isosceles (b) χ = 60° Scalene
(c) χ = 44° Right-angled
(d) χ = 60° Equilateral
Q2: (a) p = 50° (b) p = 40° (c) p = 40° (d) p= 128°
Q3: a = 65° Isosceles triangle b = 65° Angles on line and angles in triangle
Q4: m = 60° Equilateral and vertically opposite
n = 150° External angles of triangle

Quadrilaterals – page 60

Q1: (a) χ = 70° parallelogram (b) χ = 50° kite
(c) χ = 75° trapezium (d) χ = 110° rhombus
Q2: (a) Rhombus (b) Rectangle (c) Square
Q3: a+b+c = 180° d+e+f = 180° a+b+c+d+e+f =360°

Q4: (a) p=56° (b) p=107° q=95° (c) p=60 (d) p=60° q=30°
Q5: (a) 108° (b) χ=65°
Q6: χ=30°

Irregular & Regular Polygons – page 61

Q1: (a) 60° (b) 115°
Q2: (a) 360° (b) 360°
Q3: (a) 4 sides (b) 360°
Q4: (a) 6 sides (b) 360°
Q5: 74°
Q6: (a) 6 sides (b) 4 sides (c) 4 sides

Irregular & Regular Polygons (cont) – page 62

Q7: (a) 72° (b) 108°
Q8: (a) 60° (b) 120°
Q9: (a) a = 72° b = 54° (b) a = 60° b = 60°
Q10: No, interior plus exterior equals 180°
Q11: (a) 5 sides (b) χ = 80° (c) 80°, 100°, 20°, 30°, 130°
Q12: (a) 108° (b) 72° (c) 5
Q13: 36 sides

Congruence & Similarity – page 63

Q1: R and S
Q2: (a) χ= 6.4cm y = 72° (b) χ= 8cm y = 80°
Q3: See page 11
Q4: PQR and XYZ three angles
Q5: Angles in hexagon = 120° which will divide into 360°, so three hexagons can meet at a single point. Angles in pentagon = 108° which will not divide into 360°

Similar Figures – page 64

Q1: (a) No, all angles in each ▲ not equal (b) Yes, all angles in each ▲ equal
Q2: (a) XY = 10cm (b) BC = 3.9cm
Q3: (a) DE = 7.5cm (b) BO = 1.5cm
Q4: ABC and XYZ similar, three angles equal

Pythagoras' Theorem 1 & 2 – page 65

Q1: (a) 13cm (b) 25cm (c) 7.2cm (d) 16cm (e) 10.4cm (f) 16.4cm
Q2: Triangle C
Q3: 9.7cm

Pythagoras' Theorem 1 & 2 (cont) – page 66

Q4: (a) 6cm (b) 7.3cm
Q5: 13.3km
Q6: (a) 102.5cm (b) 11.2cm (c) 10.0cm
Q7: Yes, $6.25^2 - 6^2 = 1.75^2$ or $6^2 + 1.75^2 = 6.25^2$ etc.
Q8: 8.5cm
Q9: 9.5cm
Q10: (a) 4.2m (b) 3.2m
Q11: 11.4cm

Trigonometry 1, 2 & 3 – page 67

Q1: (a) 5cm (b) 11.3cm (3 s.f.) (c) 13.2cm (1d.p.)
Q2: (a) 60° (b) 41.8° (c) 45°
Q3: Triangle B

Trigonometry 1, 2 & 3 (cont) – page 68

Q4: (a) 7.2m (b) 38.9°
Q5: 5.77cm
Q6: (a) 64.1° (b) 63.4° (c) 43.1°

Q7

Q7: No, $\tan^{-1}\left(\dfrac{12.6}{6.3}\right)$ does not equal 60°
Q8: (a) 33.6m (b) 2.7m
Q9: height = 28cm or 0.28m width = 24cm or 0.24m

Transformations 1 – page 69

Q1: (a) (i) L (ii) C (iii) D (iv) J (v) G (vi) H (b) Reflection in line y = 2.5
(c) Reflection in line χ = 1
Q2: A, Reflection in line χ = -2 B, Reflection in line y = χ
C, Reflection in line y = -χ D, Reflection in line y = 0.5
Q3: See page 11

Transformations 2 – page 70

Q1: (a) (i) F (ii) H (iii) C (iv) I (v) E
(b) 90° clockwise about origin (c) 90° anticlockwise about (-2,0)
Q2: See page 11
Q3: See page 11

Transformations 3 – page 71

Q1: (a) (i) D (ii) H (iii) E (iv) C (v) I (vi) J (b) $\binom{-4}{-2}$ (c) $\binom{8}{-4}$
Q2: See page 11
Q3: See page 11

Transformations 4 – page 72

Q1: (a) (b) (c) (d) See page 11
Q2: (a) centre (-1,0) Scalefactor 2
(b) centre (15,1) Scalefactor $^1/_2$
Q3: See page 11

Transformations 5 – page 73

Q1: (a) (i) Reflection in line y axis (ii) Reflection in χ axis
(b) Rotation 180° about origin
Q2: (a) (i) Rotation clockwise 270° about origin (ii) Reflection in χ axis
(b) Reflection in line y = χ
Q3: (a) (b) See page 11 (c) Rotation 180° about (1,1)
Q4: (a) (b) (c) and (d) See page 11 (e) Rotation 180° about (2,0)
(f) Reflection in line y = χ

Coordinates 1 – page 74

Q1: A(1,3) B(2,-1) C(4,2) D(-4,3) E(-4,-1) F(-2,-4) G(5,-3) H(0,1)
Q2: (a) P(1,2) Q(2,4) R(5,4) (b) S(4,2) (c) (i) PQ=(1.5,3) (ii) QR=(3.5,4)
(iii) RS=(4.5,3) (iv) PS=(2.5,2)

Coordinates 2 – page 75

Q1: (a) See page 11 (b) (i)(2,3$^1/_2$) (ii)(5,5$^1/_2$) (iii)(7,4$^1/_2$) (iv)(4,2$^1/_2$)
Q2: (a) See page 11 (b) (i) PQ=3.6 (ii) QR=5 (iii) RS=3.6 (iv) SP=5
Q3: (2,1$^1/_2$)
Q4: (a) See page 12 (b)(i)(-3,1$^1/_2$) (ii)(3,1$^1/_2$) (iii)(3,-2) (iv)(-3,-2)
(c) AB=DE=4cm AF=BC=3.6cm EF=CD=4.5cm Perimeter=24.2cm
Q5: See page 12

Perimeter 1 & 2 – page 76

Q1: (a) 23.2cm (b) 22.1cm (c) 22.5cm (d) 33.1 (e) 16cm (f) 36cm

Perimeter 1 & 2 (cont) – page 77

Q2: (a) 25.1cm (b) 30.1cm
Q3: (a) 13cm (b) 26.12cm
Q4: 6.4cm
Q5: 45.1cm
Q6: 6×10^2cm
Q7: 3.72×10^4m
Q8: (a) 75.4cm (b) 7.5cm (c) 37.7cm (d) 377.0cm

Area 1 & 2 – page 78

Q1: (a) 14cm^2 (b) 13cm^2
Q2: (a) 36cm^2 (b) 10800cm^2 or 1.08m^2 (c) 75cm^2 (d) 32cm^2 (e) 18cm^2
(f) 264cm^2 (g) 3550cm^2 (h) 314.2cm^2 (i) 78.5cm^2

Area 1 & 2 (cont) – page 79

Q3: r = 1.9m
Q4: r = 1.5m
Q5: (a) 44cm^2 (b) 0.8m^2
Q6: (a) 184cm^2 (b) 105.0cm^2
Q7: 29m^2 area cost £19.96 4 tins needed

Volume 1 & 2 – page 80

Q1: (a) 160cm^3 (b) 216cm^3 (c) 30cm^3
Q2: (a) 81cm^3 (b) 785cm^3 (c) 942cm^3
Q3: (a) 5cm (b) 5cm (c) 6.25cm

Volume 1 & 2 (cont) – page 81

Q4: (a) 25cm (b) 2cm

Q5: 39.5cm

Q6: (a) Perimeter (b) Perimeter (c) Area (d) Area (e) Volume (f) Perimeter (g) Area (h) Volume

Q7: (a) 628cm^3 (b) 6.4cm^3

Q8: (i) $\dfrac{\chi^2}{y}$ (ii) $\chi^2 + y^2$ (iii) χy^2

3-D Shapes 1 & 2 – page 82

Q1: (a) Rectangle (b) AB,EF,GH (c) DE,CF, BG (d) 8

Q2: (a) 12 (b) 7 (c) 7

Q3: See page 12

Q4: See page 12

3-D Shapes 1 & 2 (cont) – page 83

Q5: (a) (b) (c) See page 12

Q6: (a) Cuboid (b) 8 (c) See page 12 (d) 3

Q7: (a) and (d) are correct nets

Q8: All sides 3cm. Hexagons can be anywhere as long as one on top and one on bottom. See page 12

Q9: 5

Q10: (a) (b) (c) See page 12

Symmetry 1 & 2 – page 84

Q1: See page 12

Q2: See page 12

Q3: (a) 4 (b) 2 (c) 0 (d) 2 (e) 1 (f) 0

Q4: See page 12

Q5: (a) order 6 (b) order 8 (c) order 5 (d) order 3

Symmetry 1 & 2 (cont) – page 85

Q6: (a) See page 12 (b) See page 12

Q7: (a) (b) See page 12 (c) 5

Q8: (a) See page 12 (b) See page 12, order 6

Q9: (a) 4 (b) 2 (c) 2 (d) 2 (e) 1 (f) 1

Scale Drawings & Map Scales – page 86

Q1: (a) See page 12, all measurements in cm (b) 29.25m

Q2: (a) 60km (b) 24km (c) Bodmin, Holsworthy

Q3: See page 12, scale 1cm - 10m, 94m

Q4: (a) See page 12 (b) 6.3m

Measuring Bearings – page 87

Q1: (a) (i) 090° (ii) 047° (iii) 125° (iv) 225° (v) 270° (vi) 305°

(b) (i) 35km (ii) 50km (iii) 30km

Q2: A ⟶ B 045°

B ⟶ C 145°

C ⟶ D 215°

D ⟶ A 308°

Q3: (a) See page 13 (b) 312°

Q4: See page 13

Converting Measurements – page 88

Q1: (a) 4.5m (b) 3500ml (c) 1250g (d) 6.874kg (e) 45000m (f) 5.5mm

Q2: (a) 10.05m (b) 1937m (c) 2.65m

Q3: 4220mg, 0.405kg, 420g, 4kg, 39.5kg

Q4: (a) 18 inches (b) 180g (c) 22.5litres

Q5: Jake by 20cm

Q6: 1200m, 1050000mm, 1km, 900m, 11000cm

Q7: (a) 4.8km (b) 7.5 miles (c) 2025g (d) 0.8 pounds

(e) 6.9litres (f) 39.4 pints

Q8: 11000 yards

Compound Measures – page 89

Q1: 1hour 40mins Q2: 11.2mph Q3: 24m Q4: total distance = 370km, total time = 4 hours, average speed = 92.5km/h Q5: 107.1kg/m^3

Q6: vol = 500cm^3, change by 50cm^3

Q7: (a) 108km/hr (b) 378km

Q8: (a) 0.8g/cm^3 (b) 35200g or 35.2kg

Constructions 1 & 2 – page 90

Q1- Q5: See page 13

Constructions 1 & 2 (cont) – page 91

Q6: See page 13

Q7: (a) See page 13 (b) All angle bisectors intersect in middle

Q8: Correct construction

Q9: Correct constructions

Loci – page 92

Q1- Q4: See page 13

Q5: (a) See page 13 (b) $y=\chi$ (c) See page 13 (d) $y=2\chi$

Properties of Circles 1 & 2 – page 93

Q1: A centre, B Radius, C Diameter, D Circumference, E Chord, F Tangent

Q2: See page 13

Q3: (a) 45° (b) 67.5°

Q4: (a) See page 13

(b) OTX and OWX OTV and OWV TVX and WVX (c) 140°

Properties of Circles 1 & 2 (cont) – page 94

Q5: (a) isosceles (b)(i) 20° (ii) 20° (c)(i) 70° (ii) 70° (iii) 40° (d) isosceles

Q6: (a) Correct construction

(b) Triangle OAD is isosceles, OA=OD, radius of circle is perpendicular bisector of AD hence AX = XD

Q7: OA=10cm=radius, Area of circle=100π or 314cm^2

Q8: (a) (i) OY=OW (both radii) OX=OX (same line) $O\hat{X}Y=O\hat{X}W=90°$
∴ Trangles OXY and OXW are congruent ∴ WX=YX and so OX bisects WY (ii) OX=6.2cm (1 d.p.)

(b) AB and AC are parallel, OB, OD, OA, OC are all radii hence equal.
△ OAC is congruent to △ OBD Hence AC= BD

Properties of Circles 3 – page 95

Q1: (a) 124° Angles subtended by an arc.

(b) 102° Angles subtended by an arc.

(c) 90° Angles subtended by a semicircle.

Q2: (a) $y = 30°$ Angle subtended by a semicircle

$\chi=30°$ Isoceles triangle

(b) $\chi = 40°$ Opposite angles in a cyclic quadrilateral.

$y = 140°$ Opposite angles in a cylic quadrilateral. (c) $\chi=90°$, $y=45°$

Q3: (a) AEC = 68° Opposite angles in a cyclic quadrilateral.

DEC= 52°, EDC = 90° angles in a semicircle.

Hence AED = 68° + 52° = 120°

(b) AED = 110° (ABC = 110° angles straight line, AEC = 70° opposite angles in a cyclic quadrilateral)

Q4: (a) 20° (b) 19° (c) 33°

Q5: Kite 2, 120°+60°=180°. Sum of opposite angles=180°

HANDLING DATA

Probability – page 96

Q1: (a) (i) $^4/_{10} = ^2/_5$ (ii) $^3/_{10}$ (iii) $^2/_{10}=^1/_5$ (iv) $^1/_{10}$ (v) $^6/_{10} = ^3/_5$ (vi) $^7/_{10}$

(b) 0 _____ 0.5 _____ 1

D C B A

Q2: (a) $^{28}/_{50}$ (b) $^{22}/_{50}$ (c) 0.44<0.7 Harder at Centre A

Q3: (a) $^1/_2$ (b) $^1/_2$ (c) $^{10}/_{30} = ^1/_3$ (d) $^{22}/_{30} = ^{11}/_{15}$ (e) $^8/_{30} = ^4/_{15}$

Mutually Exclusive Outcomes – page 97

Q4: (a) 4 diamonds + 2 diamonds, 4 diamonds + 2 hearts, 2 spades + 2 diamonds, 2 spades + 2 hearts, 5 spades + 2 diamonds, 5 spades + 2 hearts

(b) (i) 2/6 = 1/3 (ii) 1/6 (iii) 5/6

Q5: 0.35

Q6: (a) $^8/_{15}$ (b) $^1/_{15}$

Q7: (a) Disagree - probability>1 (b) Agree - 18 chewy, 30 sweets

(c) Disagree - Probability still $^1/_2$

Q8: (a) $^7/_{10}$ (b) $^3/_{10}$ (c) $^3/_{10}$ (d) $^7/_{10}$ (e)1 (f) 0

Q9: (a) (i) $^3/_{12}=^1/_4$ (ii) $^6/_{12}=^1/_2$ (iii) $^2/_{12} = ^1/_6$ (b) (i) $^3/_4$ (ii) $^1/_2$ (iii) $^5/_6$ (c) 24

Listing all Outcomes – page 98

Q1: (a)

2	3	4	5	6	7
3	4	5	6	7	8
4	5	6	7	8	9
5	6	7	8	9	10
6	7	8	9	10	11
7	8	9	10	11	12

(b) (i) $6/36 = 1/6$ (ii) $15/36 = 5/12$ (iii) $15/36 = 5/12$ (iv) $7/36$ (v) 0
(vi) $12/36 = 1/3$ (vii) $12/36 = 1/3$
Q2: (a)

	1	1	2	10	20	20
1	2	2	3	11	21	21
2	3	3	4	12	22	22
5	6	6	7	15	25	25
5	6	6	7	15	25	25
10	11	11	12	20	30	30
50	51	51	52	60	70	70

(b) (i) $4/36 = 1/9$ (ii) $3/36 = 1/12$ (iii) $12/36 = 1/3$ (iv) $6/36 = 1/6$ (v) $5/6$ (vi) 0
Q3: (a)

	1	2	3	4	5	6
2	2	4	6	8	10	12
4	4	8	12	16	20	24
8	8	16	24	32	40	48
9	9	18	27	36	45	54
10	10	20	30	40	50	60

(b) (i) $2/30 = 1/15$ (ii) $2/30 = 1/15$ (iii) $9/30 = 3/10$ (iv) $3/30 = 1/10$
(v) $3/30 = 1/10$ (vi) $27/30 = 9/10$

Relative Frequency – page 99

Q1: (a) (i) 60 (ii) 40 (iii) 20
(b) Relative Frequency
0.55
0.32
0.13
Q2: (a) Relative Frequency
100 0.44
150 0.53
200 0.46
250 0.48
(b) See page 13
(c) 0.5 Theoretically $1/2$ landing on tails
Q3: (a) Relative Frequencies 0.1 0.08 0.11 0.13 0.15 0.16 0.17 0.18
0.17 0.16 0.16 0.16
(b) See page 13
(c) 250

Tree Diagrams 1 & 2 – page 100

Q1: (a) 0.7
0.7
0.3
0.7
0.3
0.3

(b) (i) 0.49 (ii) 0.09 (iii) 0.21 (iv) 0.21 (v) 0.42
Q2: (a) 0.2
(b)
0.5 0.3
0.2
0.5
0.5 0.3
0.3 0.2
0.2
0.5 0.3
0.2

(c) (i) 0.25
(ii) 0.09 + 0.06 + 0.06 + 0.04 = 0.25
(iii) 0.06
(iv) 0.62 (1 - 0.38) not won by same girl

Tree Diagrams 1 & 2 (cont) – page 101

Q3: (a)
$5/11$ B
$4/11$ W
$2/11$ R
$6/12$ B
$4/12$ W
$2/12$ R
$6/11$ B
$3/11$ W
$2/11$ R
$6/11$ B
$4/11$ W
$1/11$ R

(b) (i) $1/3$ (ii) $2/3$ (iii) $5/22$
Q4: (a)
0.8 Y
0.8 Y
0.2 N
0.8 Y
0.2 N
0.2 N

(b) (i) 0.64 (ii) 0.32 (iii) 0.04

Q5: (a) Sat Sun
$1/2$ W
$1/6$ D
$1/3$ L
$1/2$ W
$1/6$ D
$1/3$ L
$1/2$ W
$1/6$ D
$1/3$ L
$1/2$ W
$1/6$ D
$1/3$ L
$1/2$ W
$1/6$ D
$1/3$ L

(b) (i) $1/4$ (ii) $1/4$ (iii) $1/2$ (iv) $4/9$
Q6: (a)
R
$3/11$ B
$3/11$ G
$2/11$ $3/11$ Y
R
$1/3$ R $3/11$ B
$3/11$ G
$1/4$ B $2/11$ $3/11$ Y
$1/4$ G R
$1/6$ Y $3/11$ B
$3/11$ G
$2/11$ $3/11$ Y
R
$3/11$ B
$3/11$ G
$2/11$ $3/11$ Y

(b) (i) $13/66$ (ii) $53/66$ (iii) $14/33$

Collecting Data 1 & 2 – page 102

Q1: Primary data is obtained firsthand usually by you. Secondary data is obtained from an external agency eg. data from internet.
Q2: (a) Choosing a reliable, balanced group of people, ages, sex, etc.
(b) No, should be taken from all over UK, males and females.
Q3: (a) Not suitable - leading question
(b) Suitable - establishes if they eat breakfast
(c) Suitable - may not have time for breakfast
(d) Not suitable - not relevant
(e) Suitable - establishes age group
(f) Suitable - establishes what they eat
(g) Suitable - differences
(h) Not suitable - not relevant

Collecting Data 1 & 2 (cont) – page 103

Q4: (a) How much do you spend?
Less than £20 ☐ £20-£40 ☐ £40-£60 ☐ £60-£80 ☐
How often do you shop here?
Less than once a week ☐ Once a week ☐ More than once a week ☐
(b) Any variations - Do you agree we are the best supermarket?

Q5: (a)

	Tally	Frequency
Less than 1 hour		
1 - 2 hours		
2 - 3 hours		
3 - 4 hours		
4 - 5 hours		
more than 5 hours		

(b) Ask an equal number of males and females from different age groups. A high street would be a good place to conduct this survey (or somewhere similar).
Q6: (a) Any variations.
(b) Any variations.

Q7: (a)

	Tally	Frequency
Sweets		
Books		
Magazines		
Clothes		
Presents		
Music		
Games		
Other		

(b) Collect data during breaktime, in assembly, etc.

Sorting Data 1 & 2 – page 104

Q1: Discrete is "counted" e.g. whole numbers
Continuous is "measured" e.g. time, weight etc.

Q2:
1	IIII	4
2	JHH II	7
3	JHH III	8
4	JHH JHH II	12
5	JHH IIII	9
		40

Q3: (a)
1	III	3
2	JHH	5
3	JHH III	8
4	JHH JHH III	13
5	JHH IIII	9
6	II	2
		40

(b) 80%

Sorting Data 1 & 2 (cont) – page 105

Q4:
0-19	II	2
20-39	III	3
40-59	JHH	5
60-79	JHH IIII	9
80-99	I	1
		20

Q5: (a)
5≤ T<10	III	3
10≤T<15	JHH JHH II	12
15≤T<20	JHH JHH II	12
20≤T<25	III	3
		30

(b) 50%

Q6:
Goals	Freq
0	11
1	19
2	5
3	0
4	3
	38

Q7: (a)
160≤h<165	1
165≤h<170	4
170≤h<175	8
175≤h<180	8
180≤h<185	4
185≤h<190	3
190≤h<195	2
	30

(b) 83$\frac{1}{3}$%

Q8:
50≤w<55	II	2
55≤w<60	I	1
60≤w<65	JHH II	7
65≤w<70	III	3
70≤w<75	JHH II	7
75≤w<80	JHH III	8
80≤w<85	JHH III	8
85≤w<90	III	3
90≤w<95	I	1
		40

Sorting Data 3 – page 106

Q1: (a)
0	8 9
1	0 7
2	1 3 8 9
3	0 1 4 4 8
4	1 1 2 4 5 6 7 7
5	1 2 3 5 6 7 8
6	0 8
7	4

(b) 41$\frac{1}{2}$ (c) 74-8=66

Q2: 32 [41] 73
[23] 24 [47]
55 [65] [120]

Q3: (a) 42 [30] 18 [90]
[16] 12 [13] 41
14 [24] 31 69
[72] [66] 62 [200]

(b) 45%

Q4:
(a)
16	4 6 7 9 9
17	0 1 1 3 3 3 4 4 5 5 6 8 8 8 8 9
18	0 1 2 4 6 8 9
19	1 2

(b) 170≤h<180
(c) 175$\frac{1}{2}$

Displaying Data 1 – page 107

Q1: See page 13
Q2: See page 13
Q3: (a) Skimmed (b) 1100 pints (c) See page 14
Q4: (a) and (b) See page 14
Q5: (a) and (b) See page 14

Displaying Data 2 & 3 – page 108

Q1: See page 14
Q2: (a) See page 14 (b) See page 14
(c) Women generally spend more on cosmetics than men.
Modal amount women 4≤m<6 Modal amount men 2≤m<4

Displaying Data 2 & 3 (cont) – page 109

Q3: (a) See page 14 (b) Ages 1 and 2
Q4: Seasonal trend Q5: 1st 88, 2nd 78, 3rd 71, 4th 84
Q6: (a)
15≤ T<20	JHH JHH I	11
20≤T<25	JHH JHH III	13
25≤T<30	JHH IIII	9
30≤T<35	JHH II	7
		40

(b) See page 14 (c) See page 14

Scatter Diagrams 1 & 2 – page 110

Q1: (a) The longer the journey the greater the distance travelled
(positive correlation) (b) See page 15
(c) (i) approx. 42.5 minutes (ii) approx. 1.6km
Q2: (a) See page 15
(b) Positive correlation - as height increases weight increases
(c) (i) approx. 137$\frac{1}{2}$ m (ii) approx. 86kg

Scatter Diagrams 1 & 2 (cont) – page 111

Q3: (a) See page 15
(b) Negative, as change received increases number of items decreases
(c) (i) £4.40 (ii) 6 items
Q4: (a) See page 15 (b) (i) 61 mins (ii) 8 tracks
Q5: (a) See page 15
(b) The more chapters the more pages - positive
(c) (i) 295 pages (ii) 15 chapters

Pie Charts – page 112

Q1: (a) See page 15 (b) 35% (c) $^7/_{20}$ $(^{42}/_{120})$
Q2: (a)
Grades	Frequency
A	6
B	13
C	12
D	7
E	2
	40

(b) See page 15 (c) $^{19}/_{40}$

Pie Charts (cont) – page 113

Q3: (a) (i) 24 (ii) 20 (b) 12 men, 15 women
Q4: (a) (i) National (ii) Daily Gossip (b) (i) 18 (ii) 16 more
Q5: (a) See page 15 (b) 40%
Q6: (a)
1 - 20	JHH JHH JHH III	18
21 - 40	JHH III	8
41 - 60	JHH I	6
61 - 80	JHH I	6
81 - 100	II	2
		40

(b) See page 15 (c) 35%
Q7: (a) (i) 6 (ii) 8 (b) (i) 36 (ii) 24

Mean, Median Mode & Range 1 – page 114

Q1: Mean 6, Median 2, Mode 2, Range 19
Q2: Mean £205, Median £210.50, Mode £248, Range £95
Q3: (a) Mean 56kg, Median 53, Mode 51, Range 19
(b) Average weight, lower with smaller distribution of weights

Q4: (a) Mean 1.56m, Median 1.59m, Mode 1.62m, Range 0.32m
(b) Same range of heights but boys average height 12cm bigger

Mean, Median Mode & Range 2 & 3 – page 115

Q1: (a)40 (b)1 (c)1 (d) 4 (e) 1.5

Q2: (a)24 (b)4 (c)3 (d) 3

Q3: (a)

Mid Temp	Fre x Mid Temp
7.5	22.5
12.5	175
17.5	192.5
22.5	45
	435

Mean 14.5°C
(b) $10 \le t < 15$ (c) $10 \le t < 15$

Mean, Median Mode & Range 2 & 3 (cont) – page 116

Q4: (a)

Mid Interval point m.i.p (χ)	χ x F
1	3
3	45
5	40
7	35
9	9
	132

Mean = $^{132}/_{32}$ = 4.1 (1 d.p.)

(b) $2 < m \le 4$ (c) $2 < m \le 4$

Q5: (a) 30 (b) $^{380}/_{30}$ = 12.7 (1 d.p.) mins

Q6: (a) (i) 1.8 goals (ii) 6 (iii) 2 goals (iv) 2 goals (b) year 10 by 1 goal

Q7: (a) $^{563}/_{33}$ =17 (b) $16 < m \le 18$ (c) $16 < m \le 18$

Cumulative Frequency 1 & 2 – page 117

Q1: (a) 8 (b) 22 (c) 43 mins

Q2: (a) Cumulative Frequency 2, 6, 11, 18, 30, 35

(b) See page 15

(c) (i) 20 (ii) 11 (d) 19 (e) 5

Cumulative Frequency 1 & 2 – page 118

Q3: (a) 14 (b) 10 (c) 2

(d) Cumulative Frequency 4, 17, 36, 70, 80 (e) See page 15

(f) 30 (g) Girls results much better. 10 girls got more than 40 compared with 4 boys. Only 17 girls got 20 or less compared with 60 boys.

Q4: (a) Cumulative Frequency 10, 49, 144, 269, 300

(b) See page 15 (c) (i) 160cm (ii) 12.5cm (iii) 210

Q5: (a) See page 15 (b) (i) 70kg (ii) 12 (c) $52^{1}/_{2}$%

Answers to graphs and diagrams

P43 Q1(a) (b) (c) (d)

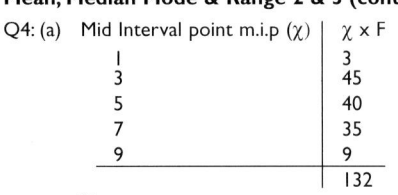

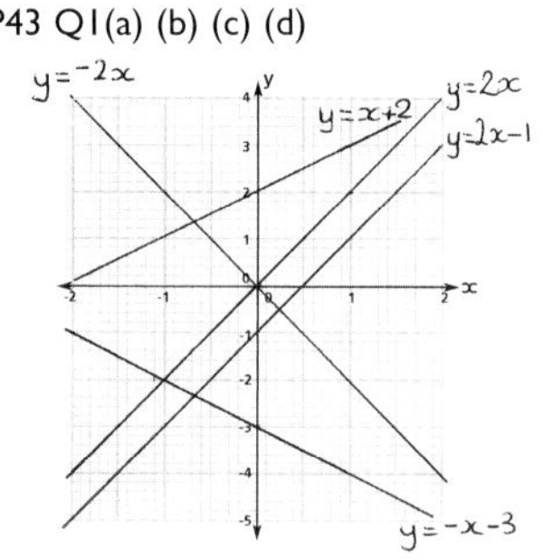

P43 Q2(b)

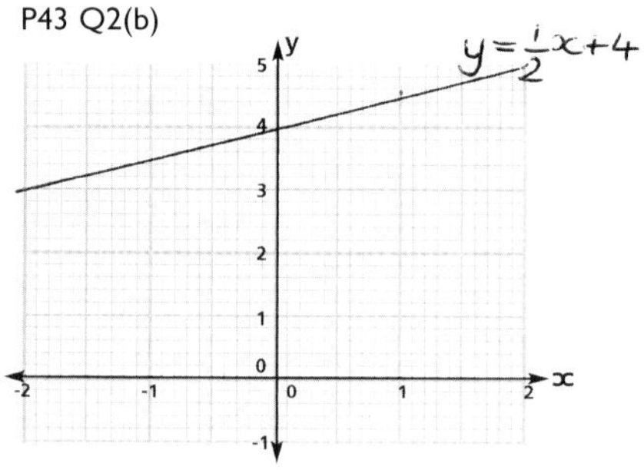

P43 Q3(b)

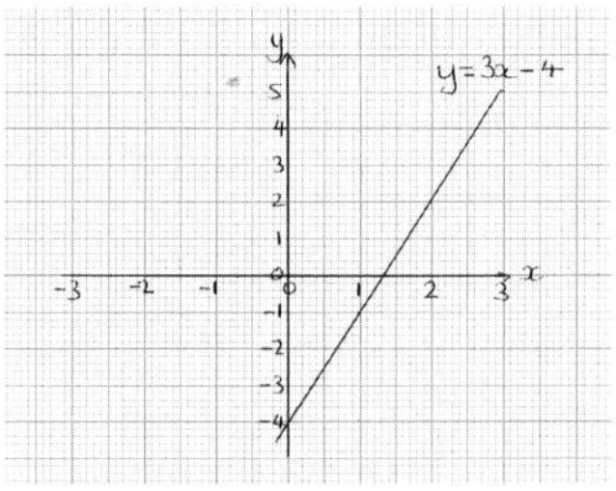

P43 Q4(a)

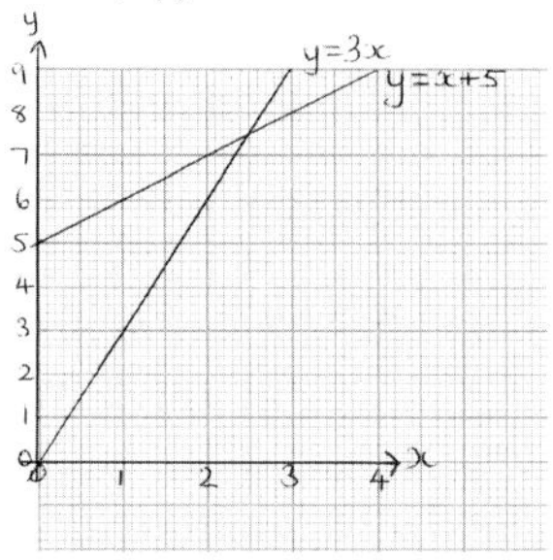

P43 Q5

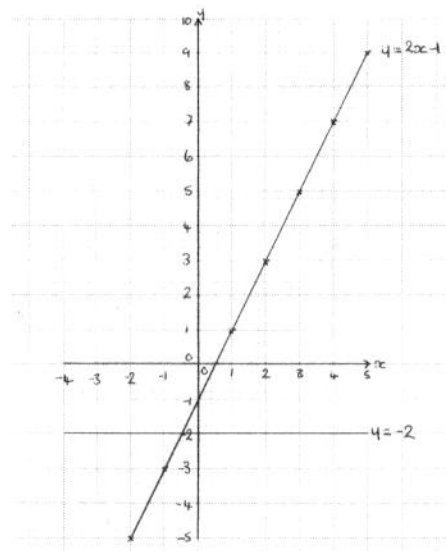

P47 Q3(a)

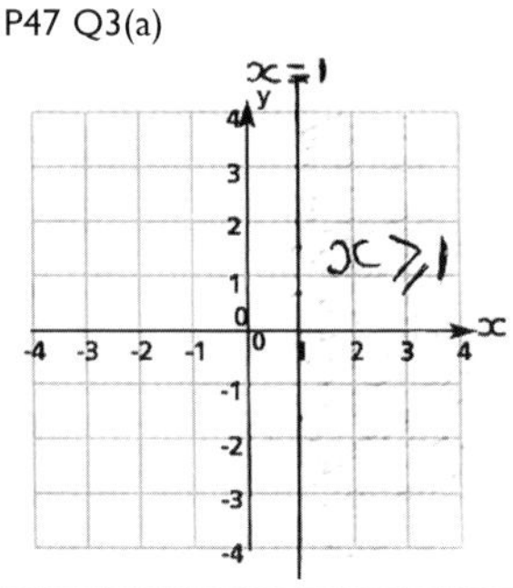

P47 Q3(b)

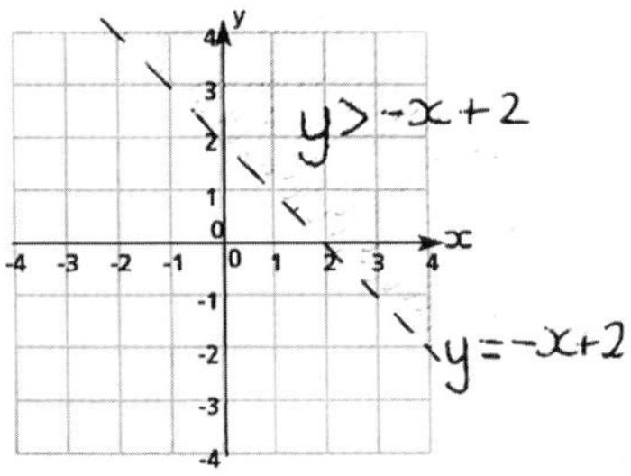

P47 Q3(c)

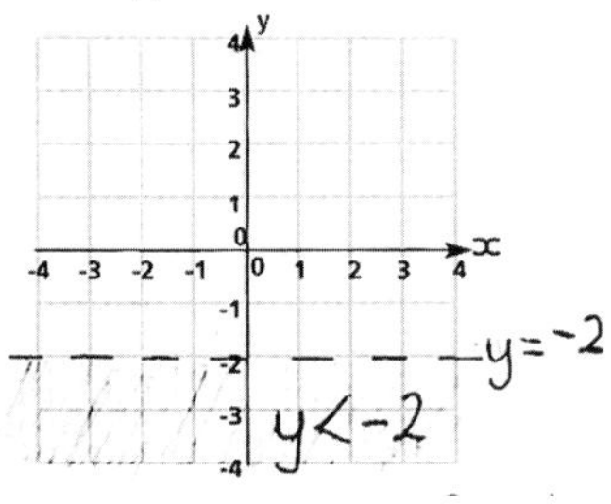

P47 Q3(d)

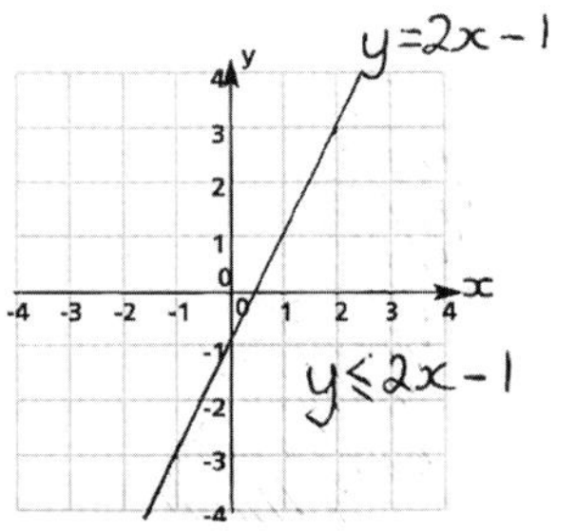

P48 Q4(a)

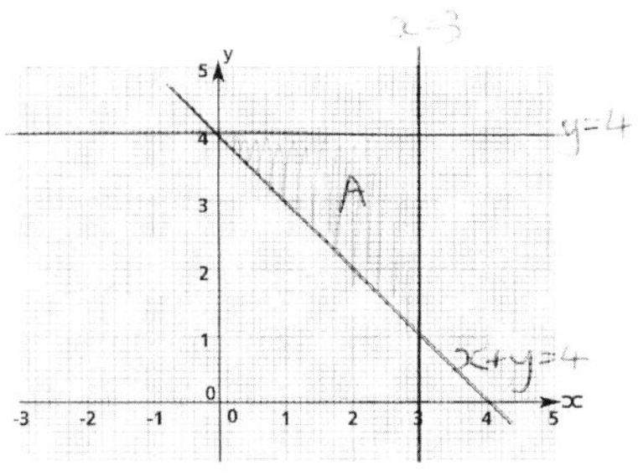

P48 Q5(a)

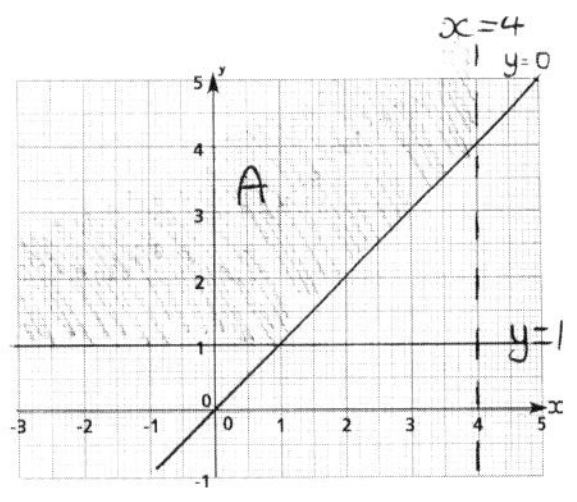

P48 Q6

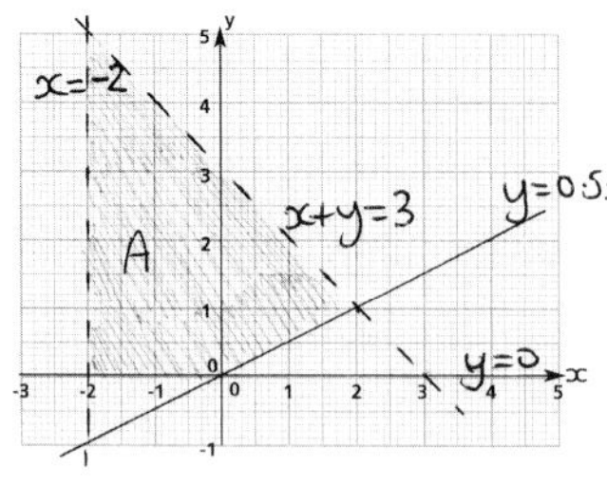

P48 Q9

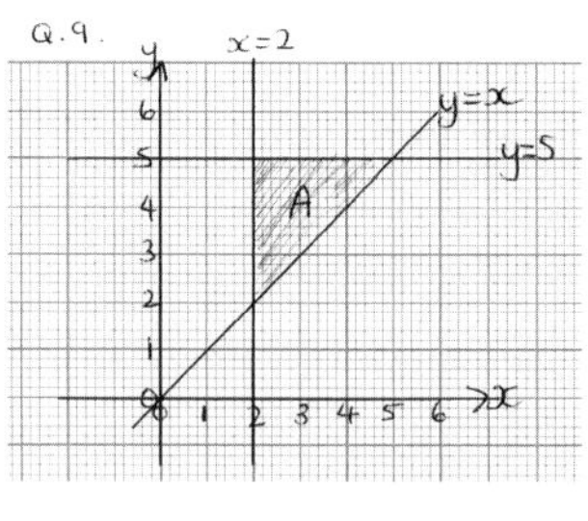

P48 Q10

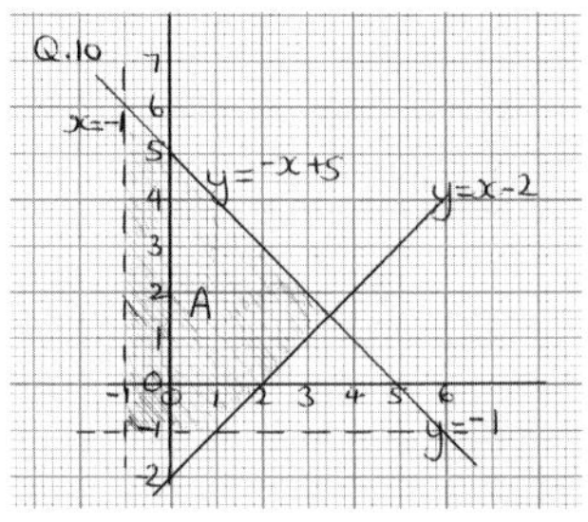

P50 Q1(a)

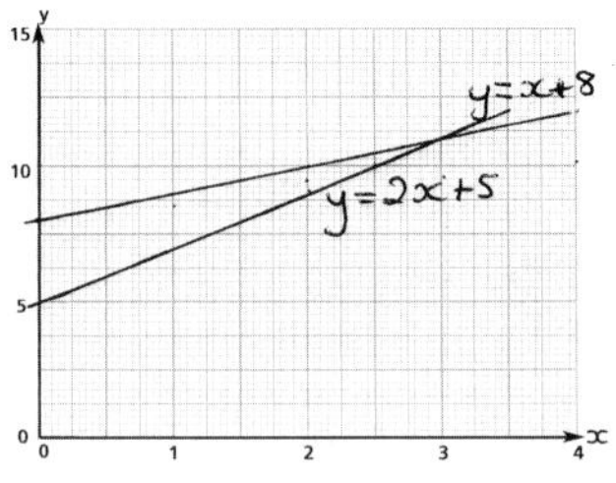

P50 Q2(a)

$y = 3x + 10$
$y = -x + 22$

P50 Q3(a)

$y = -x + 4$
$y = x + 2$
$y = 0.25x + 1$

P50 Q4(a)

$3x = y - 3$
$2x + y = 13$

P50 Q5(a)

$x + y = 6$
$3x - y = 2$
$x - y = 4$

P51 Q1(a)

b)(ii)
(b)(i)

P51 Q2(b)

c(ii)
c(i)
e
f

P52 Q3(b)

P52 Q4(b)

c(ii)
c(iii)

P52 Q5(a)

b(ii)
b(iii)

P52 Q6(a)

b(iii)
b(ii)

P53 Q1(b)

c(ii)
c(i)

P53 Q2(a)

b(i)
b(ii)

P53 Q3(b)

d(i)
d(ii)

P54 Q2(b)

Distance (km)
Time (mins)

P54 Q3(a)

Velocity (metres per second, m/s)
Time (s)

P54 Q4

Speed (m/s)
Time (secs)

P54 Q5

TRAIN B
TRAIN A
Time (hours)
1pm 2pm 3pm 4pm 5pm

10

P55 Q1(b)

Cost (£) vs Miles Travelled

P55 Q3(b)

US Dollars ($) vs Pounds Sterling (£)

P55 Q4(a)

Cost (£) vs Time (hours)

P55 Q5

Depth (cm) vs Time (mins)

P63 Q3

P69 Q3(a) and (b)

P70 Q2(a) (b) (c) (d) and (e)

P70 Q3

P71 Q2(a) (b) (c) and (d)

P71 Q3(a) and (b)

P72 Q1(a) (b) (c) and (d)

P72 Q3(a) and (b)

P73 Q3(a) (b)

P73 Q4(a) (b) (c) and (d)

P75 Q1(a)

P75 Q2(a)

11

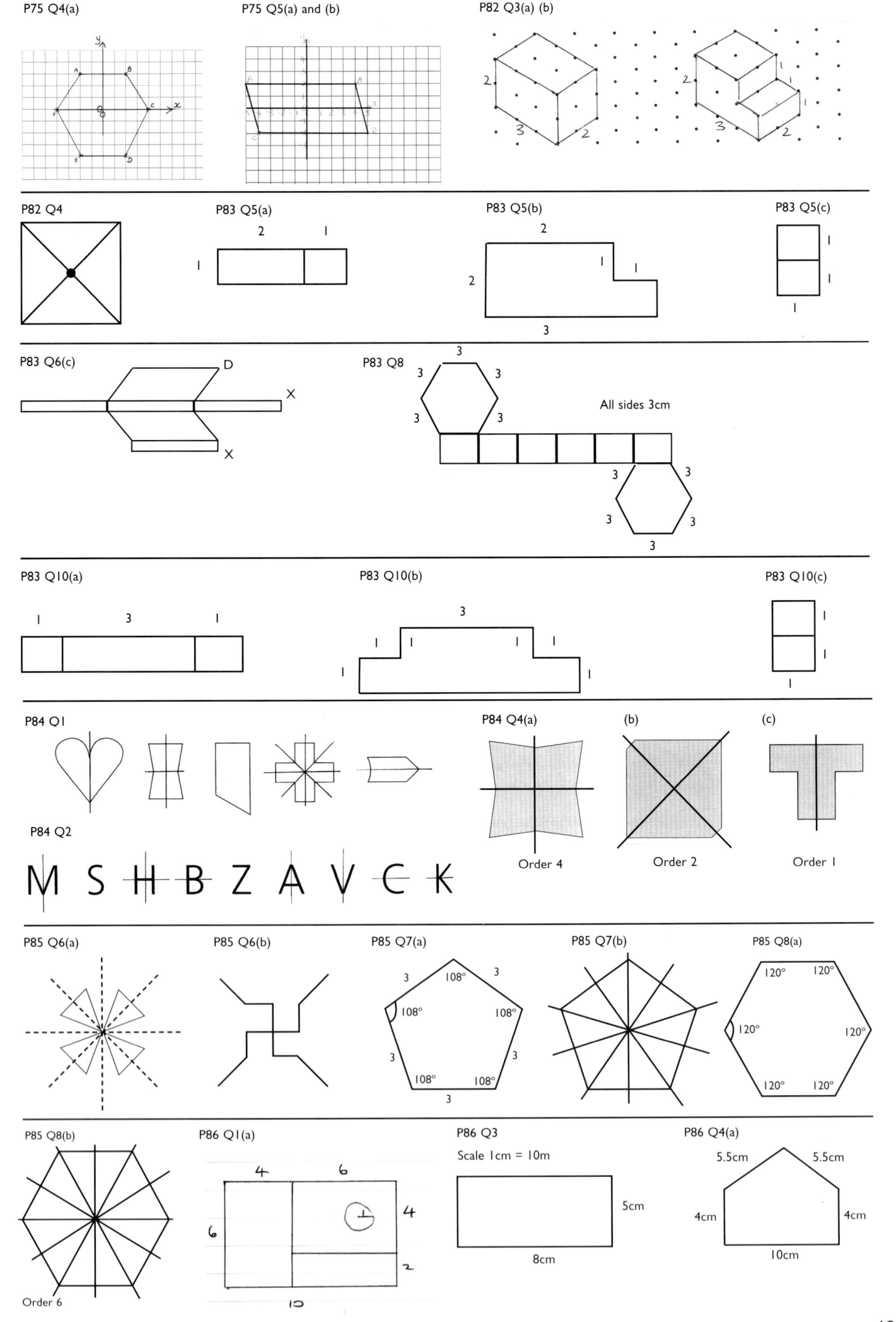

P75 Q4(a)

P75 Q5(a) and (b)

P82 Q3(a) (b)

P82 Q4

P83 Q5(a)

P83 Q5(b)

P83 Q5(c)

P83 Q6(c)

P83 Q8

All sides 3cm

P83 Q10(a)

P83 Q10(b)

P83 Q10(c)

P84 Q1

P84 Q4(a)

(b)

(c)

Order 4

Order 2

Order 1

P84 Q2

M S H B Z A V C K

P85 Q6(a)

P85 Q6(b)

P85 Q7(a)

P85 Q7(b)

P85 Q8(a)

P85 Q8(b)

Order 6

P86 Q1(a)

P86 Q3

Scale 1cm = 10m

P86 Q4(a)

12

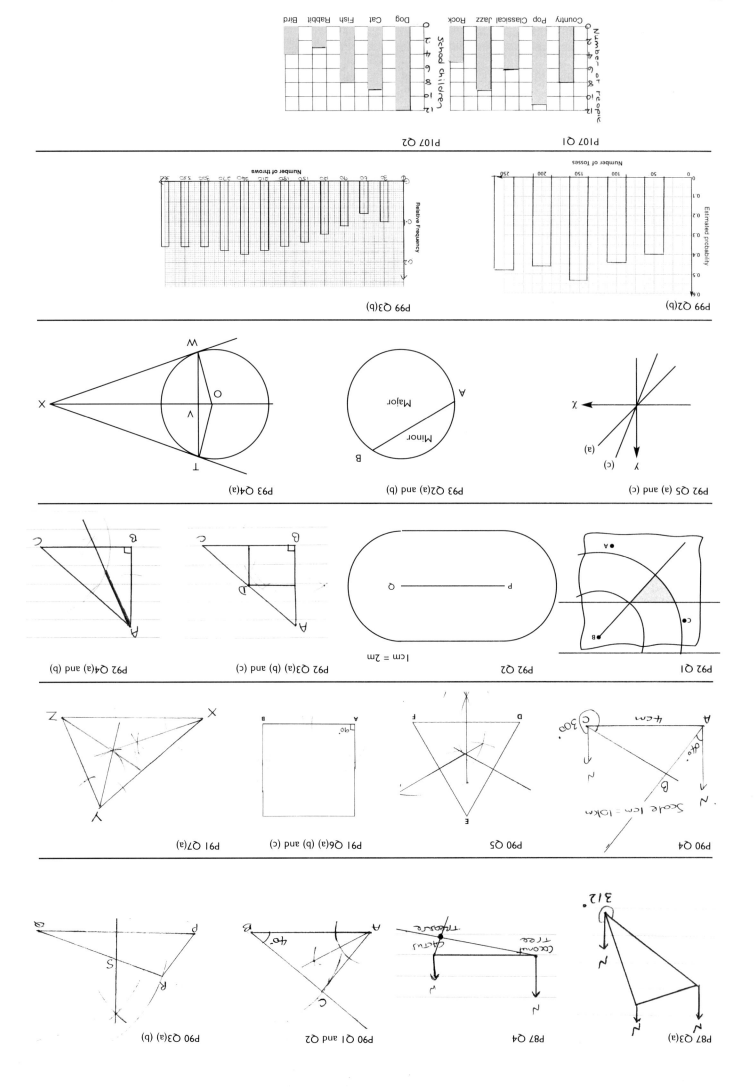

P107 Q2

P107 Q1

P99 Q3(b)

P99 Q2(b)

P93 Q4(a)

P93 Q2(a) and (b)

P92 Q5 (a) and (c)

P92 Q4(a) and (b)

P92 Q3(a) (b) and (c)

P92 Q2

P92 Q1

P91 Q7(a)

P91 Q6(a) (b) and (c)

P90 Q5

P90 Q4

P90 Q3(a) (b)

P90 Q1 and Q2

P87 Q4

P87 Q3(a)

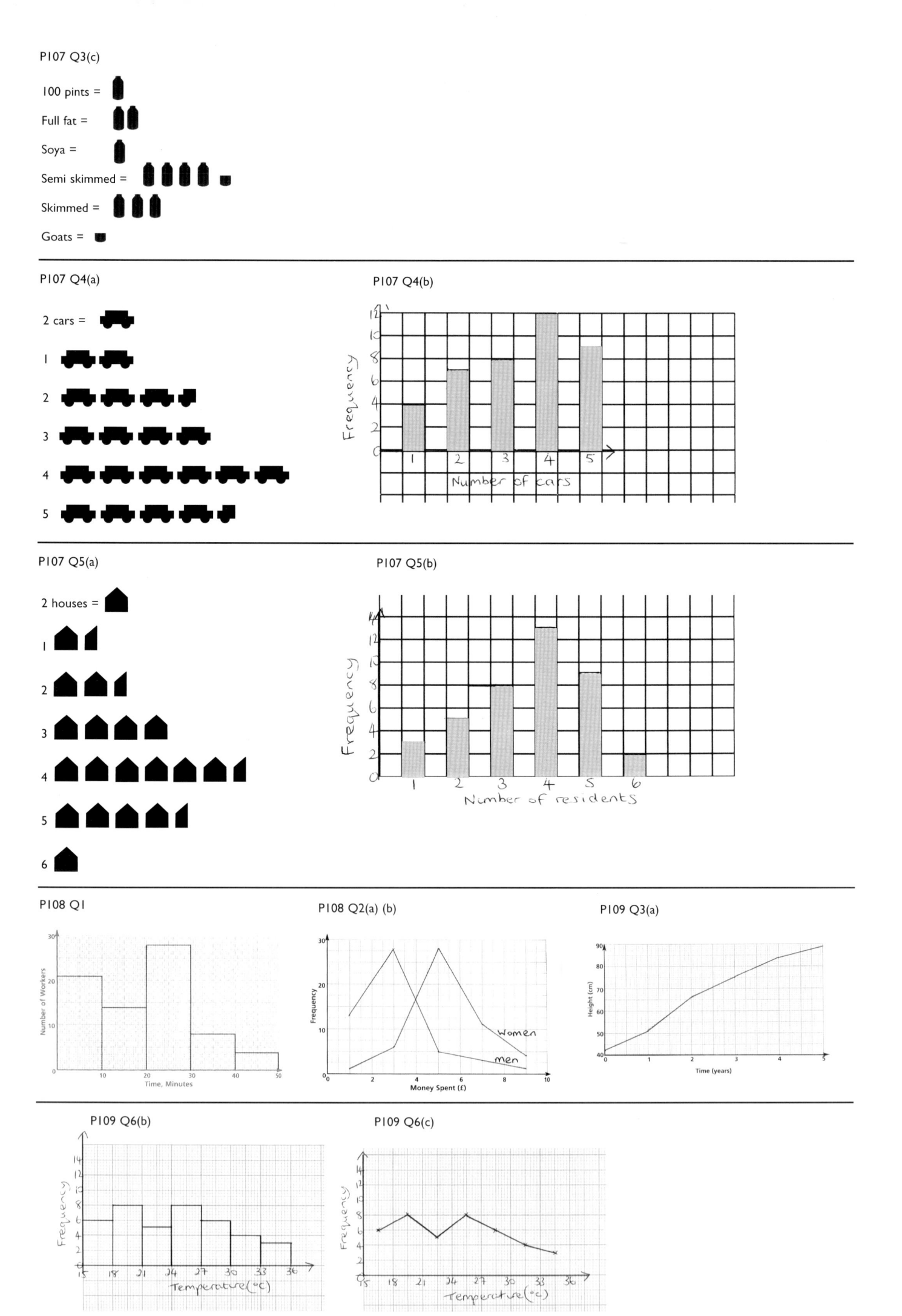

P107 Q3(c)

100 pints =

Full fat =

Soya =

Semi skimmed =

Skimmed =

Goats =

P107 Q4(a)

2 cars =

1

2

3

4

5

P107 Q4(b)

Number of cars

P107 Q5(a)

2 houses =

1

2

3

4

5

6

P107 Q5(b)

Number of residents

P108 Q1

P108 Q2(a) (b)

Women

Men

P109 Q3(a)

P109 Q6(b)

Temperature (°C)

P109 Q6(c)

Temperature (°C)

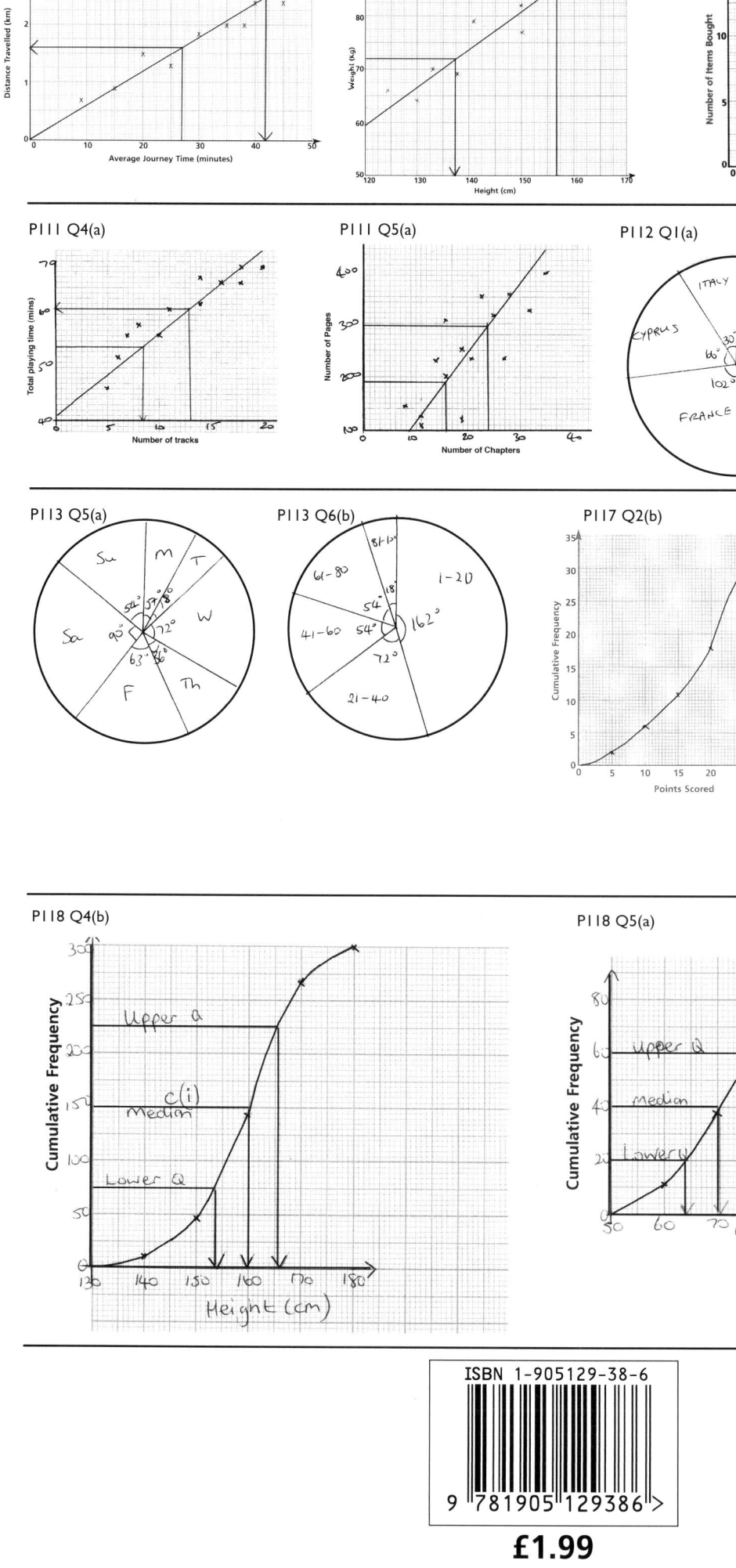